MATH KNOTS
Joy of Learning....

My Book

This book belongs to

Cover Design by :
Gowri Vemuri

First Edition :
December , 2019

Author :
Gowri Vemuri

Edited by :
Raksha Pothapragada
Ritvik Pothapragada

Questions: mathknots.help@gmail.com

NOTE : VDOE is neither affiliated nor sponsors or endorses this product.

Dedication

This book is dedicated to:

My Mom, who is my best critic, guide and supporter.

To what I am today, and what I am going to become tomorrow,

is all because of your blessings, unconditional affection and support.

This book is dedicated to the

strongest women of my life ,

my dearest mom

and

to all those moms in this universe.

G.V.

The Virginia Board of Education and Virginia Department of Education (VDOE) have developed the Virginia Assessment Program (VAP) to measure and evaluate students' academic progress in the Standards of Learning (SOLs). The SOLs indicate Virginia's expectations for what students should know and be able to do in the subject areas of reading, writing, mathematics, science, and history/social science.

Students in grades 3-12 to take the Standards of Learning (SOL) assessments each year. Some of the tests are required by all students each year, and others are required only at specific grade levels. Additionally, with the removal of some SOL tests in recent years, the VDOE assigned the responsibility of the creation and administration of alternate, performance - based assessments on local divisions. Student scores from these tests determine a school's and the division's state accreditation and measures progress toward meeting federal targets.

Virginia Standards of Learning (SOL) tests are generally given online unless a student has an identified and documented need to be assessed using paper,pencil format. The test question format is typically multiple choice, and each test contains some technology enhanced items.

GRADE 3-8 STANDARDS OF LEARNING(SOL) TESTS

GRADE 3	GRADE 4	GRADE 5	GRADE 6	GRADE 7	GRADE 8
	VIRGINIA STUDIES				WRITING
MATH	MATH	MATH	MATH	MATH	MATH
READING	READING	READING	READING	READING	READING
		SCIENCE			SCIENCE

NOTE : VDOE is neither affiliated nor sponsors or endorses this product.

END OF COURSE STANDARDS OF LEARNING (SOL) TESTS

GRADE 9	GRADE 10	GRADE 11
ALGEBRA I	GEOMETRY	ALGEBRA II
EARTH SCIENCE	BIOLOGY	CHEMISTRY
WORLD HISTORY I	WORLD HISTORY II	VIRGINIA & US HSTORY
		WORLD GEOGRAPHY
		ENGLISH : READING
		ENGLISH : WRITING

Any Student taking one of the courses listed here is expected to take the corresponding end-of-course SOL test. The grade levels depicted here represent grade level at which students typically participate in these courses.

NOTE : VDOE is neither affiliated nor sponsors or endorses this product.

SOL Test Scoring and Performance Reports:

Standards of Learning assessments in English reading, mathematics, science and history/social science are made up of 35-50 items or questions that measure content knowledge, scientific and mathematical processes, reasoning and critical thinking skills. English writing skills are measured with a two-part assessment that includes multiple-choice items and an essay.

Student performance is graded on a scale of 0-600 with 400 representing the minimum level of acceptable proficiency and 500 representing advanced proficiency. On English reading and mathematics tests, the Board of Education has defined three levels of student achievement: basic, proficient, and advanced, with basic describing progress towards proficiency.

Performance Achievement Levels:

- The achievement levels for grades 3-8 reading and mathematics tests are: *Pass/Advanced, Pass/Proficient, Fail/Basic,* and *Fail/Below Basic.*

- The achievement levels for science tests, history tests, and End-of-Course (EOC) tests* are: *Pass/Advanced, Pass/Proficient,* and *Fail/Does Not Meet.*

- The EOC Writing (2010 SOL) test, EOC Reading (2010 SOL) test, and EOC Algebra II (2009 SOL) test have an achievement level of *Advanced/College Path* in place of the *Pass/Advanced* achievement level.

NOTE : VDOE is neither affiliated nor sponsors or endorses this product.

INDEX

FORMULA SHEET

1. Area of a triangle

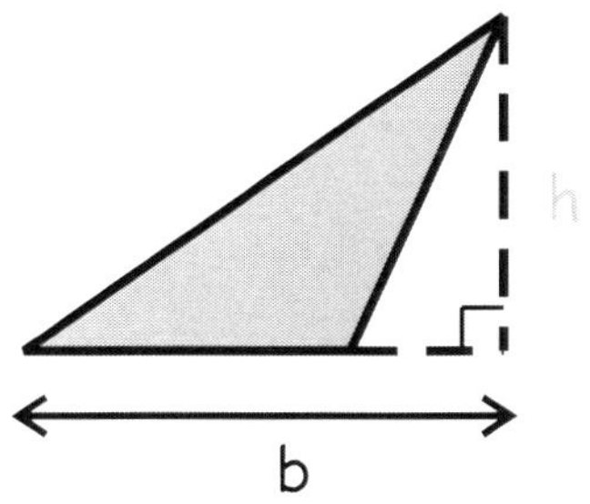

$A = \frac{1}{2} bh$

2. Area of a parellelogram

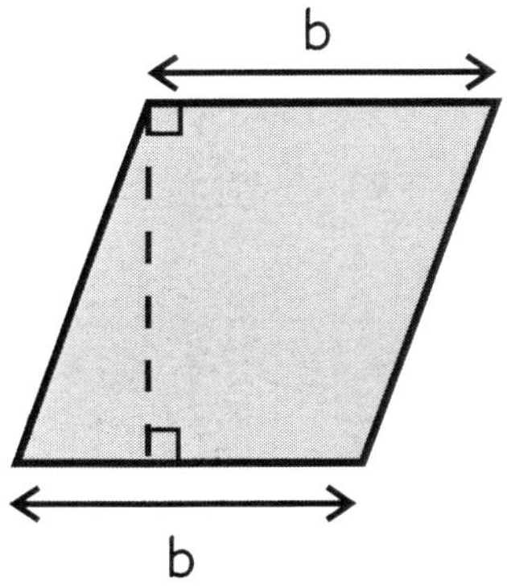

$A = bh$

3. Volume and Surface area of a cuboid

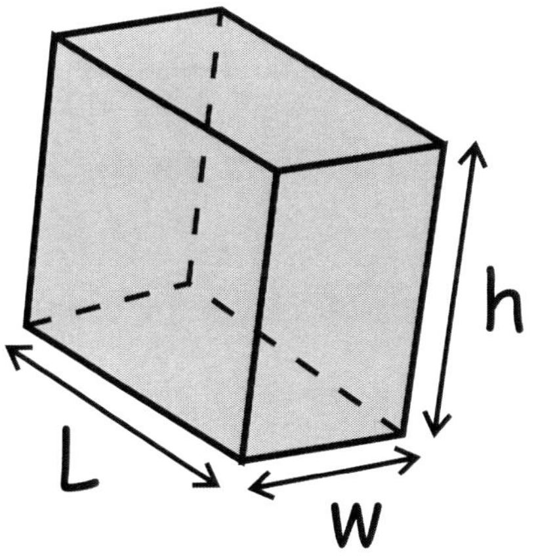

$V = lwh$

$S.A = 2(lw + lh + wh)$

4. Perimeter and Area of a Square

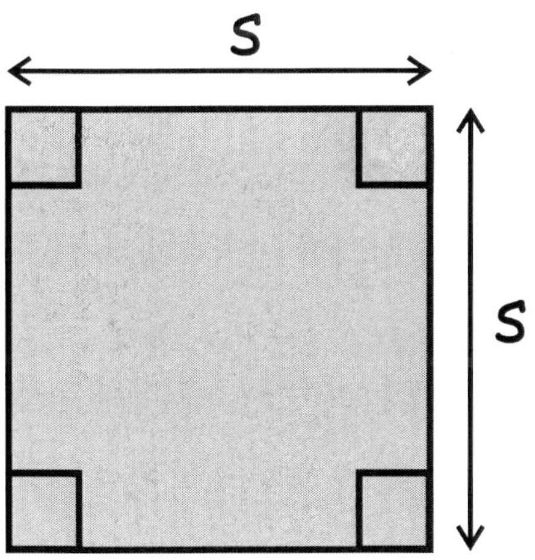

$P = 4s$

$A = s^2$

5. Area of a Trapezium

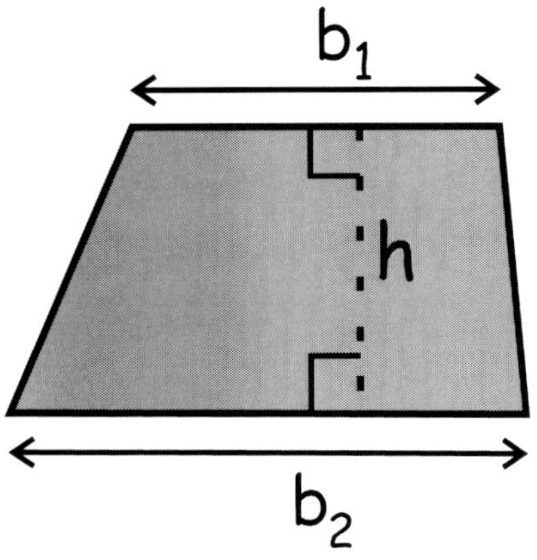

$$A = \frac{1}{2} h(b_1 + b_2)$$

6. Circumference and Area of a Circle

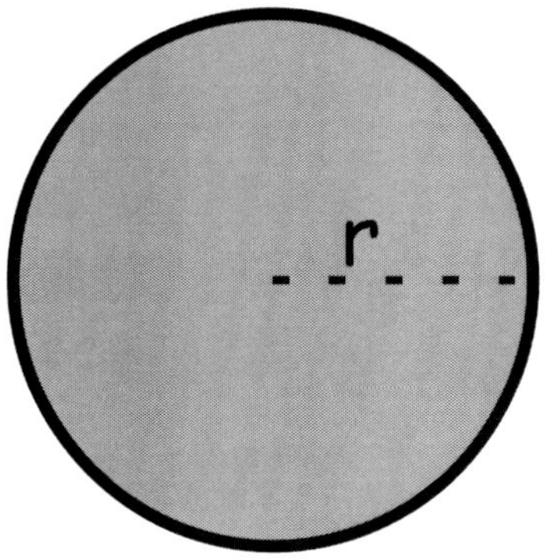

$$c = 2\pi r$$
$$A = \pi r^2$$

pi

$$\pi = 3.14$$
$$\pi = \frac{22}{7}$$

7. Perimeter and Area of a Rectangle

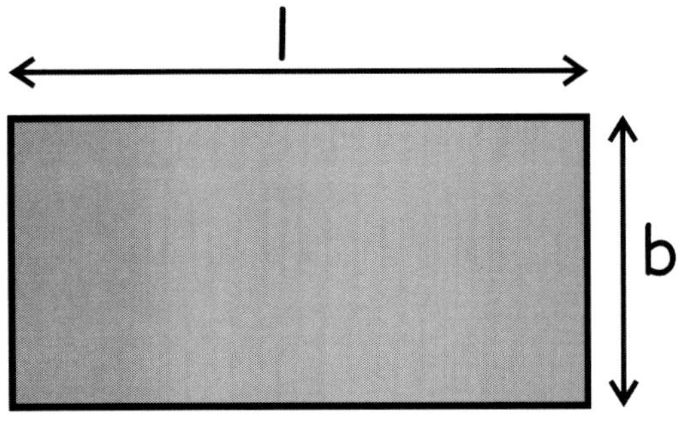

Area = l b x

Perimeter = $2(l + b)$

8. Acute angle

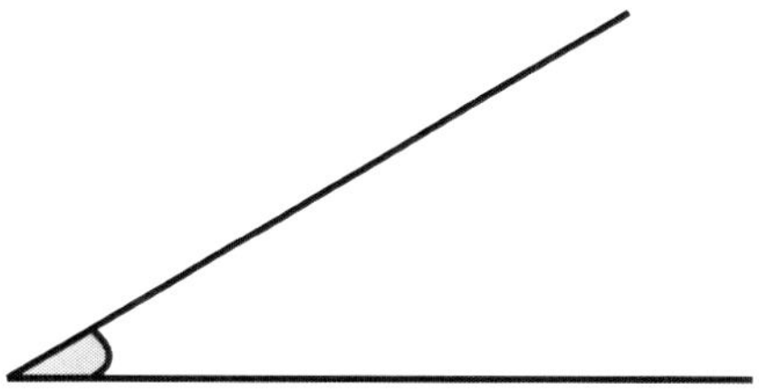

Any angle less than 90^0
is called as acute angled

9. Right angle

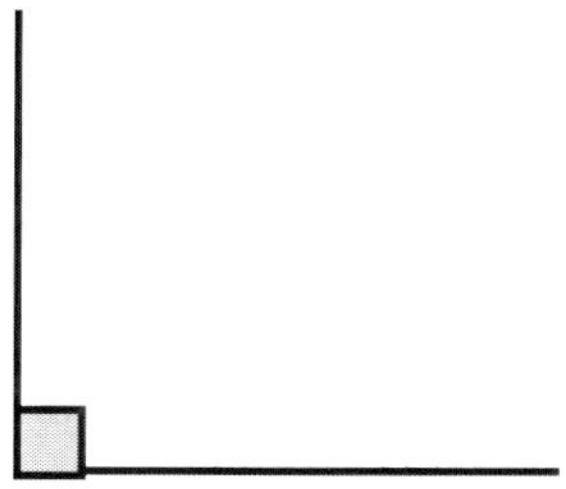

An angle equal to 90^0
is called as right angled

10. Obtuse angle

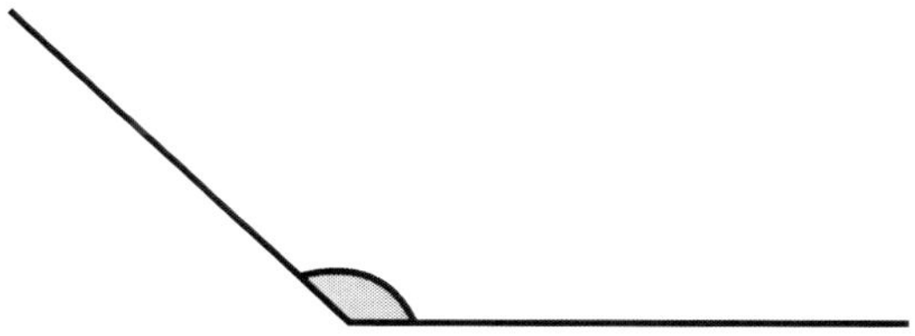

Any angle greater than 90^0
is called as obtuse angled

11. Perimeter of a polygon KLMNOP

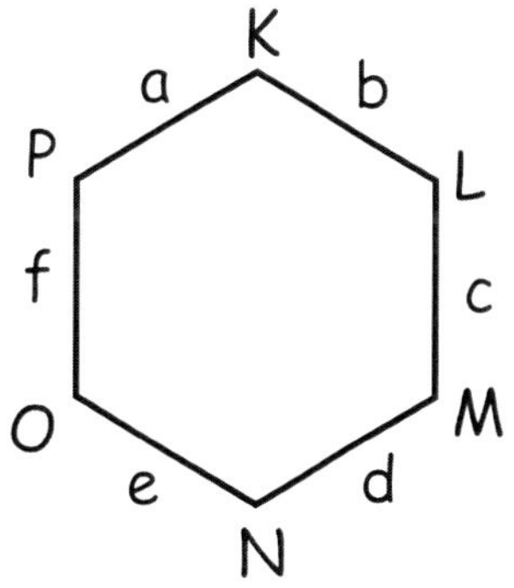

Perimeter =
(a + b + c + d + e + f) units

Abbreviations

milligram	mg
gram	g
kilogram	kg
milliliter	mL
liter	L
kiloliter	kL
millimeter	mm
centimeter	cm
meter	m
kilometer	km
square centimeter	cm^2
cubic centimeter	cm^3

year	yr
month	mon
hour	hr
minute	min
second	sec

volume	V
total Square Area	S.A
area of base	B
ounce	oz
pound	lb
quart	qt
gallon	gal.
inches	in.
foot	ft
yard	yd
mile	mi.
square inch	sq in.
square foot	sq ft
cubic inch	cu in.
cubic foot	cu ft

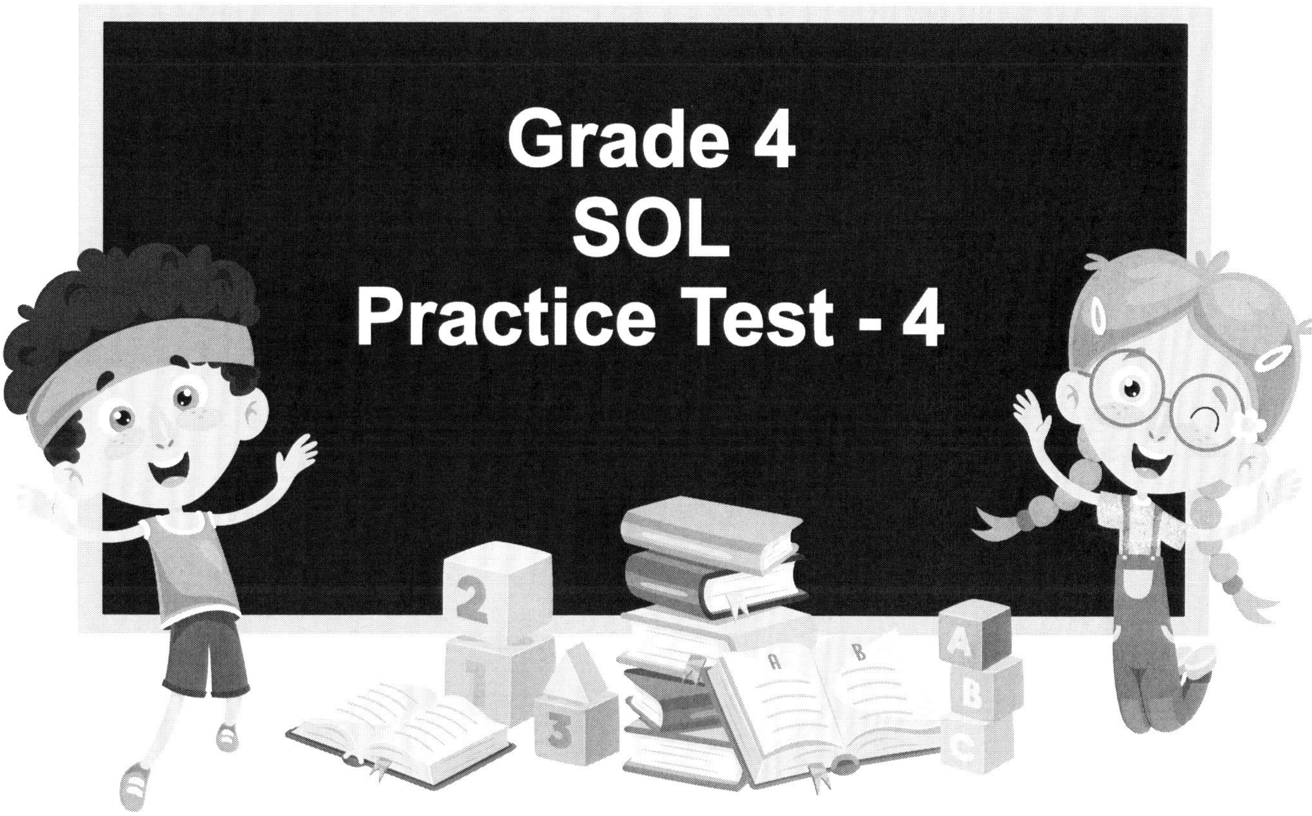
Grade 4
SOL
Practice Test - 4

1. Subtract

$$\begin{array}{r} 654{,}945 \\ -\ 37{,}829 \\ \hline \end{array}$$

(A) 617,116 (B) 617,126

(C) 618,116 (D) 616,116

2. Which of the below product is between 200 and 250 ?

(A) 11 × 11 (B) 15 × 11

(C) 23 × 11 (D) 19 × 11

3. Simplify $\frac{1}{7} + \frac{3}{7} = \boxed{?}$

(A) $\frac{4}{7}$ (B) $\frac{5}{7}$

(C) $\frac{2}{7}$ (D) $\frac{11}{7}$

4. Lola is making a fruit punch. She adds 1.83 liters of soda, 2.06 liters of grape juice, 3.22 liters of apple juice. How many liters of fruit punch was made ?

(A) 7.11 (B) 7.71

(C) 6.11 (D) 7.21

5. Below model is equal to 1.

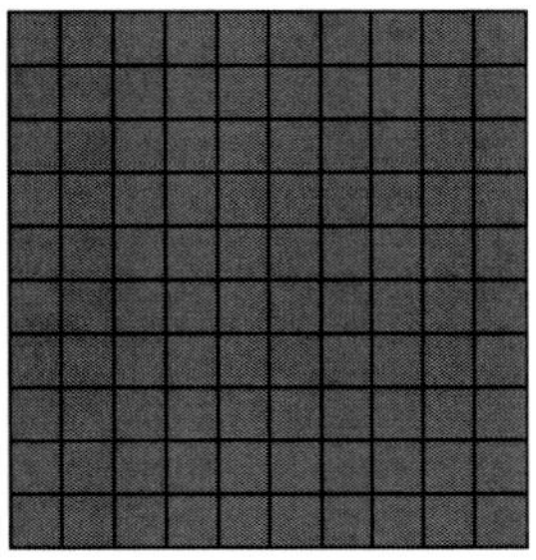

Each of the below model is shaded to represent a decimal number

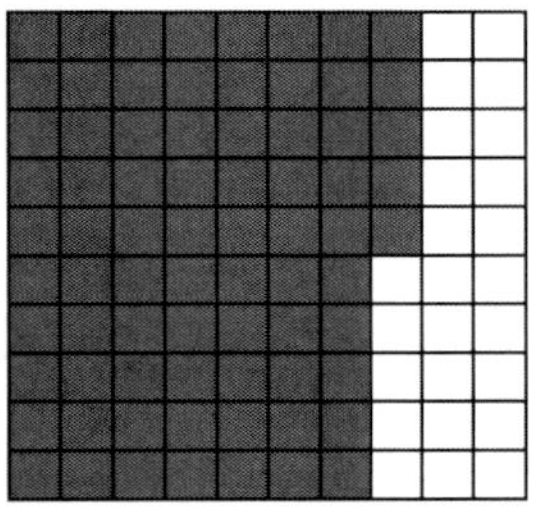

(A) 0.65 (B) 0.55

(C) 0.75 (D) 0.84

6. Which of the below statements is correct ?

(A) $\frac{77}{11} = 2 + 5$ (B) $\frac{77}{11} = 5 - 2$

(C) $\frac{77}{11} = \frac{10}{2}$ (D) $\frac{77}{11} = 77 + 11$

7. Find the next three numbers in the below pattern.

93 , 90 , 87 , 84 , 81 , 78 , ...

(A) 75, 77, 80 (B) 75, 72, 69

(C) 75, 65, 62 (D) 72, 66, 63

8. Evaluate 67.49 - 7.63 = ?

(A) 49.87
(B) 59.86
(C) 59.85
(D) 59.87

9. Simplify $\frac{908}{7}$ = ?

(A) 129 R 3
(B) 129 R 5
(C) 129 R 6
(D) 129 R 4

10. Which best describes the difference 4,927 - 2191 ?

(A) Closer to 4500 than 5000
(B) Closer to 4000 than 5000
(C) Closer to 3500 than 4000
(D) Closer to 3000 than 2000

11. Simplify $\frac{885}{2}$ = ?

(A) 442 R 0
(B) 440 R 5
(C) 442 R 1
(D) 444

12. Find the product of 63 and 13

(A) 819
(B) 818
(C) 820
(D) 817

13. Simplify $\frac{672}{7}$ = ?

(A) 95 R 6
(B) 96 R 0
(C) 96 R 1
(D) 95 R 3

14. Which two factors have a product of 1326 ?

(A) 24 × 54
(B) 56 × 26
(C) 52 × 23
(D) 51 × 26

15. Round 164,702 to the nearest thousand.

(A) 170,000
(B) 165,000
(C) 164,700
(D) 170, 702

16. Write "Seven hundred ninety-five and nine hundredths" in standard form.

(A) 795.09
(B) 7950.09
(C) 795.009
(D) 7950.009

17. Which of the following shows a fraction model and a decimal model that both represent the same value ?

(A)

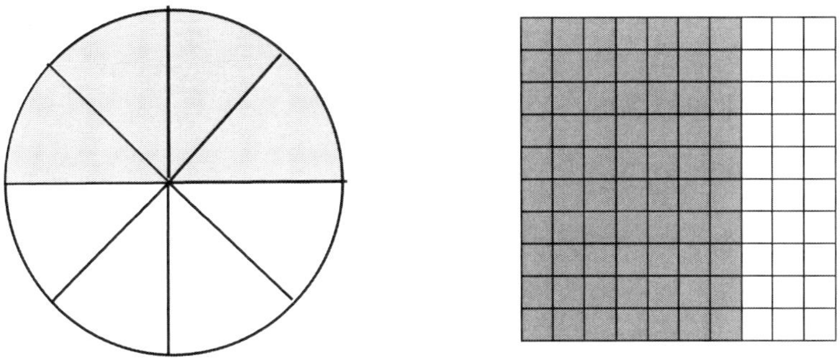

(B)

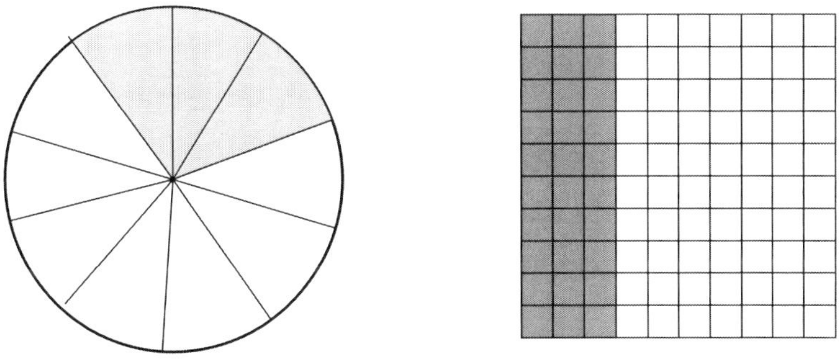

(C)

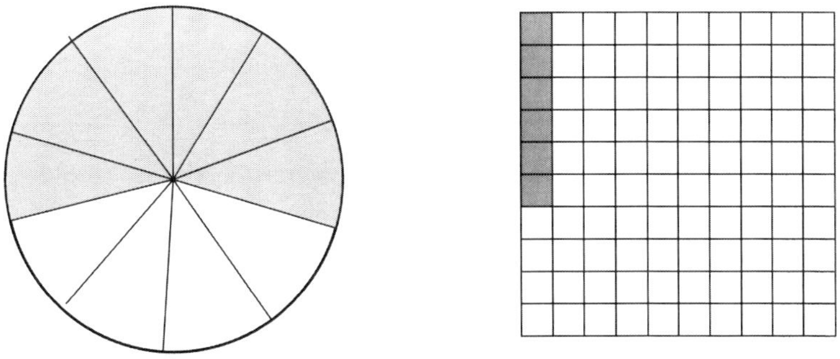

(D)

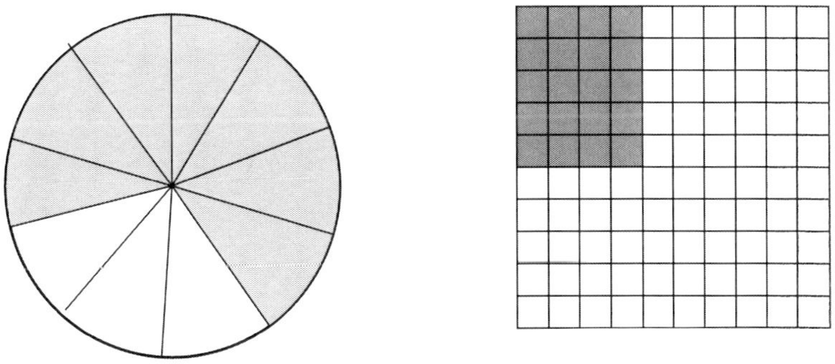

18. Which of the below model is greater than $\frac{5}{9}$

(A) 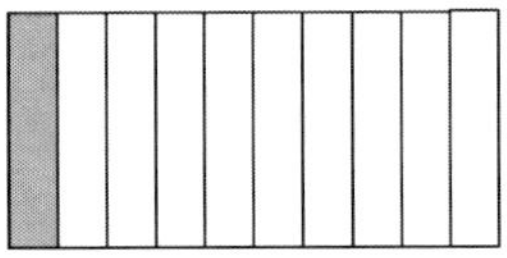

(B) 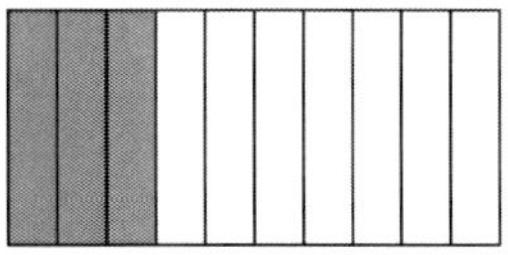

(C) 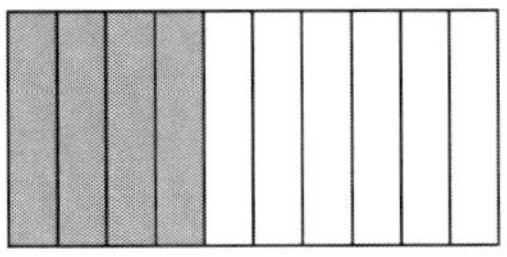

(D) 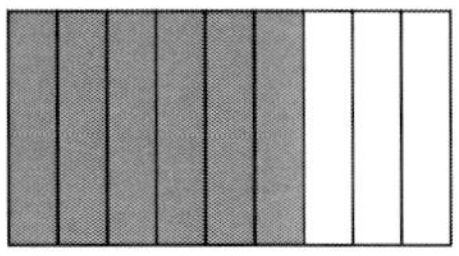

19. Which of the below is a true statement ?

(A) 262629 > 262329

(B) 262429 < 262329

(C) 362329 = 362229

(D) 262429 > 262529

20. Point X is located at $\frac{2}{5}$ on the number line

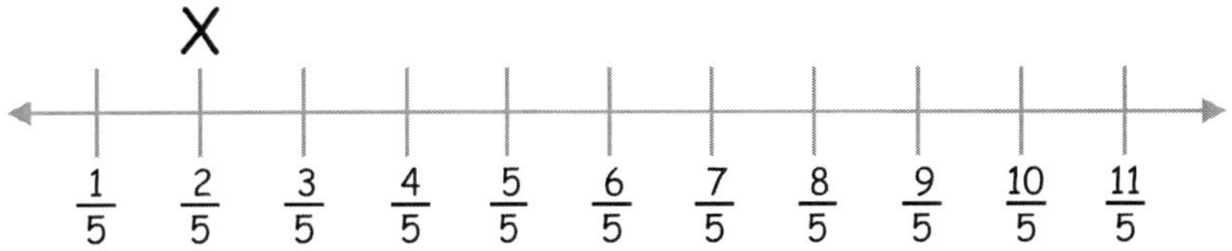

Which of the following is another name for location of X ?

(A) $\frac{4}{5}$

(B) $\frac{3}{11}$

(C) $\frac{3}{5}$

(D) $\frac{4}{10}$

21. Which of the below statements is true ?

(A) 0.654 > 0.663

(B) 0.654 < 0.644

(C) 0.654 > 0.643

(D) 0.654 = 0.650

22. The shaded sections of these models represent two decimal numbers.

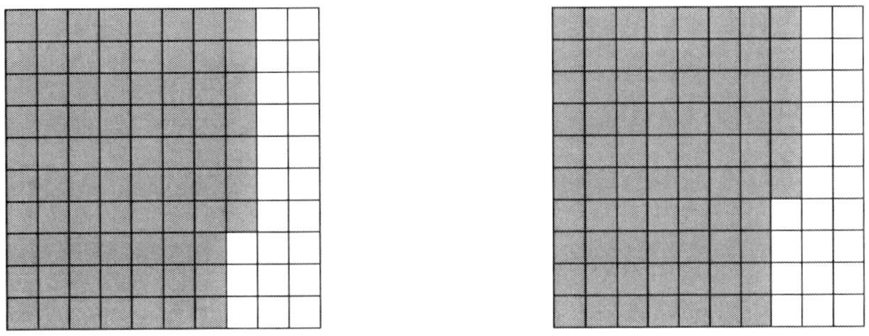

Based on the above which of the below is true ?

(A) 0.77 > 0.76

(B) 0.77 < 0.76

(C) 0.77 = 0.70

(D) 0.76 < 0.69

23. How many ml of juice is there in the jar ?

(A) 10 ml

(B) 12 ml

(C) 11 ml

(D) 9 ml

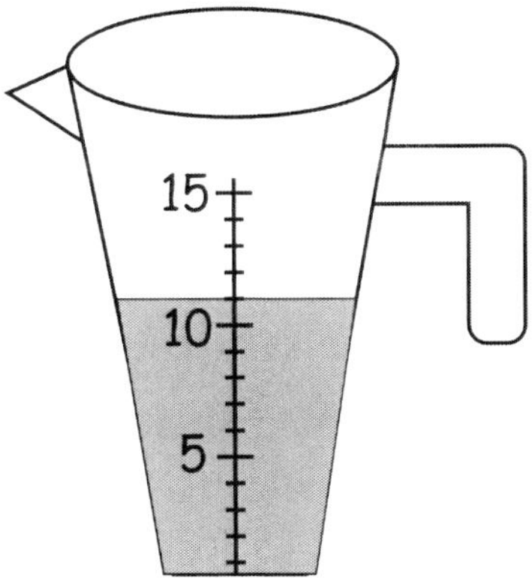

24. Which of the following is closest to the height of the bird from the ground ?

(A) 7 units

(B) 3 units

(C) 12 units

(D) 4 units

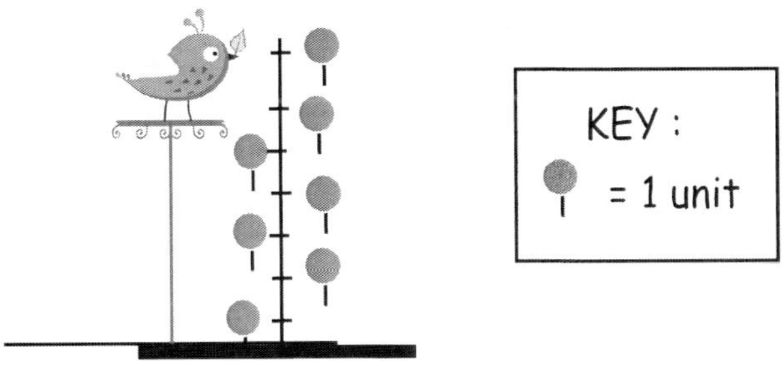

25. John is making a fruit punch with 6 pounds of orange juice. It is equal to how many ounces ?

(A) 96 ounces
(B) 84 ounces
(C) 54 ounces
(D) 92 ounces

26. Ms. Janes organic vegetable patch is 121 meters long. How long is the vegetable patch in centimeters ?

(A) 1200 cm
(B) 12100 cm
(C) 1.21 cm
(D) 121 cm

27. Which of the below is a true statement for a line ?

(A) It has one end point and extends on other way
(B) It has no end points
(C) It has a starting point
(D) It has two end points

28. Which of the below is an example of obtuse angle ?

(A)
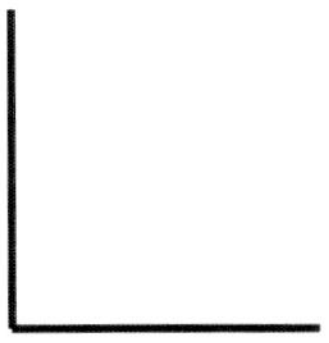

(B)
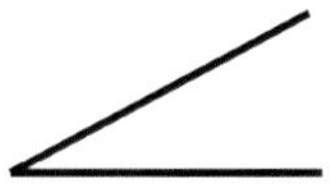

(C)
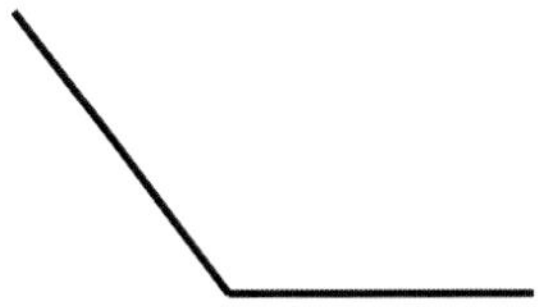

(D)

29. Which of the below are congruent.

(A) 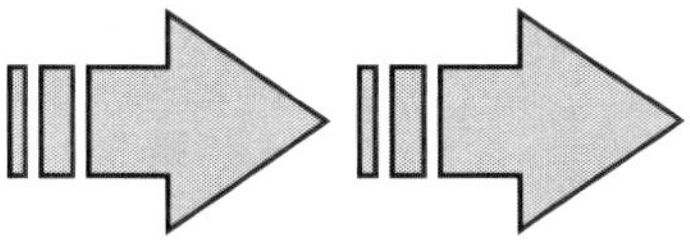

(B) 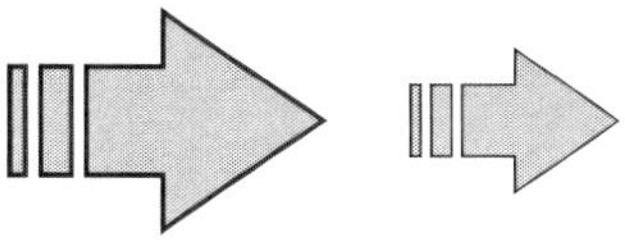

(C) 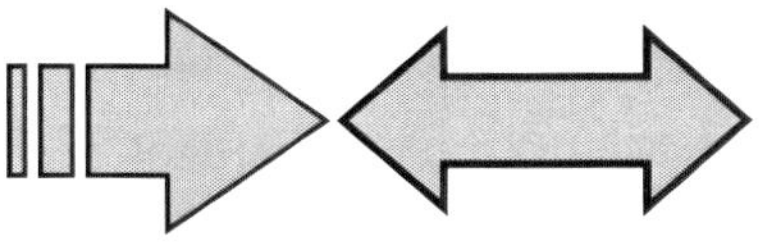

(D) 

30. Tom is baking a cake which requires 5 pints of butter. How much butter does he need in cups ?

(A) 12 cups

(B) 2.5 cups

(C) 10 cups

(D) 5 cups

31. Mason bought a cable wire of 144 inches in length. What is the cable length in feet ?

(A) 14400 feet

(B) 12 feet

(C) 1.44 feet

(D) 6 feet

32. Distance between Iris and Clara homes is 54 yards. It is equivalent to which of the below ?

(A) 162 feet

(B) 108 feet

(C) 25 meters

(D) 0.54 miles

33. Which order pair represents the missing coordinate of the below rectangle ABCD.

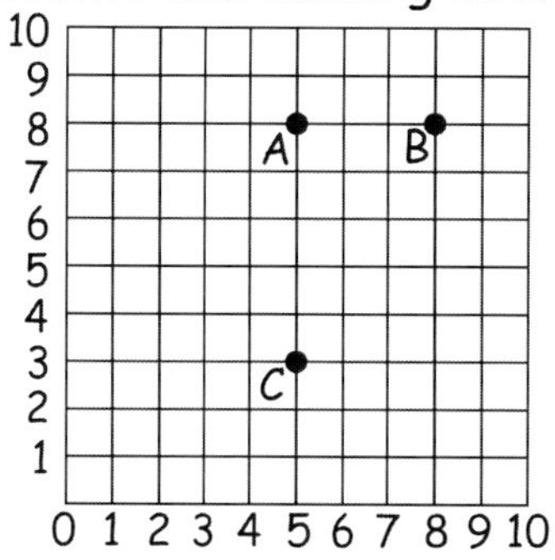

(A) (8,3)

(B) (3,3)

(C) (3,8)

(D) (8,8)

34. A magic number bag has numbered blocks as below:

#2	5 blocks
#5	8 blocks
#7	9 blocks
#11	2 blocks

Oscar picks one number block from the bag without looking in. Which color bead is least likely to be picked up ?

(A) #11

(B) #7

(C) #2

(D) #5

35. Based on the below table find the rule used on In.

In	Out
3	10
7	14
8	15

(A) Add by 2

(B) Divide by 2

(C) Add 6

(D) Add 7

36. Below graph shows the number of paintings Bella made in the last 10 days. Which of the below is the estimated total number of paintings made ?

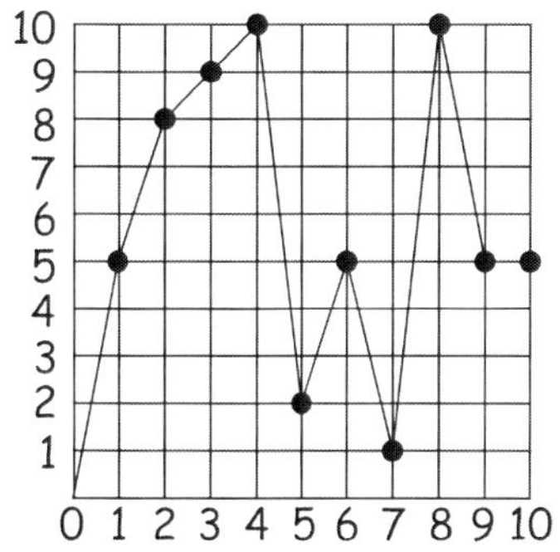

(A) 60

(B) 65

(C) 59

(D) 55

Below graph shows the number of hours Erin ran on her school track in last 8 days.

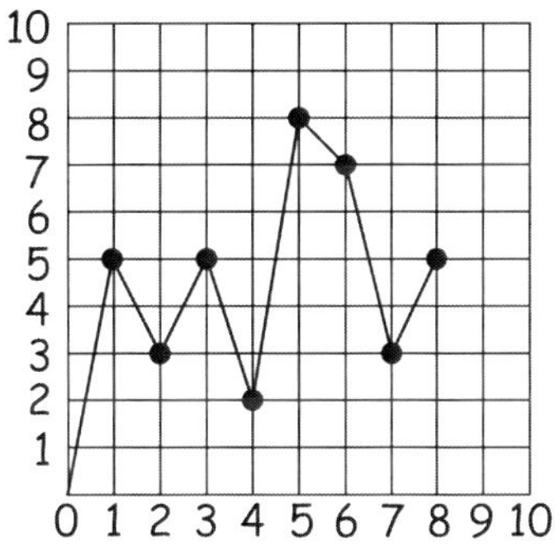

37. How many hours did she ran in 8 days ?

(A) 35 (B) 48

(C) 38 (D) 39

38. On which day she ran maximum number of hours ?

(A) Day 1 (B) Day 6

(C) Day 5 (D) Day 8

39. A number Machine throws out the below numbers

5 , 3 , 25 , 17 , 19 , 29 , 35 , 40 , 55 , 32

Molly picks a number. What is the probability that the number is a multiple of five ?

(A) $\frac{1}{2}$ (B) $\frac{4}{10}$

(C) $\frac{3}{10}$ (D) $\frac{7}{10}$

40. The table below shows various Christmas gifts that are distributed to children.

Green Sweaters	39
Red Sweaters	71
Yellow Sweaters	92
White Sweaters	11

Ayla receives one of the above. What is the chance of her getting a black sweater ?

(A) Unlikely, but not impossible (B) Impossible

(C) Likely, but not certain (D) Certain

41. Which of the below statements is correct ?

(A) $25 \times 5 = 25 + 5$ (B) $25 \times 5 = 85 - 40$

(C) $25 \times 5 = \frac{210}{2}$ (D) $25 \times 5 = 85 + 40$

42. Iris will pick one tile from a bag without looking. From which of the following bags is she is most likely to pick a tile with a "12" on it ?

(A)

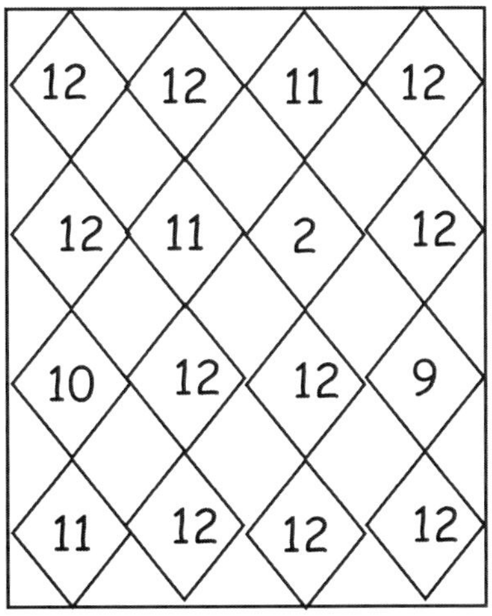

(B)

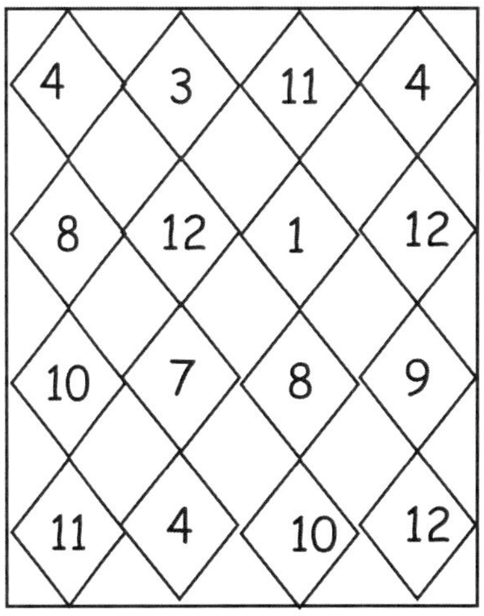

(C)

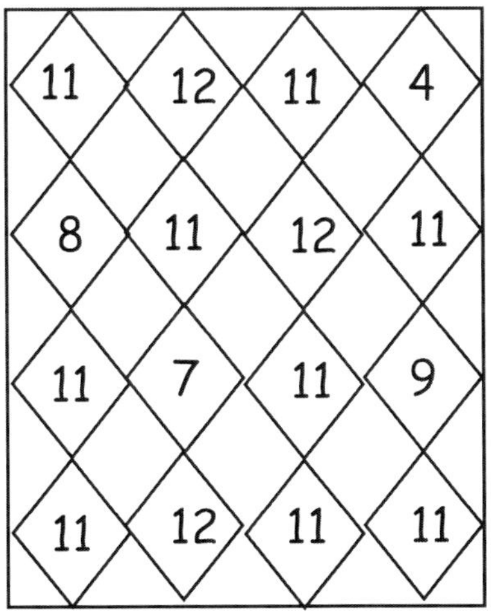

(D)

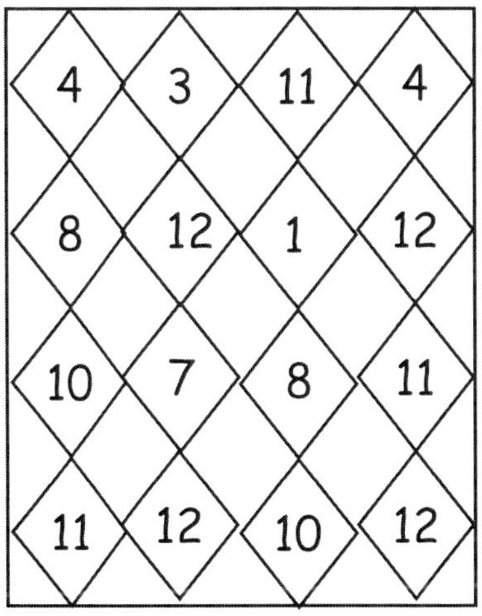

43. Freya is playing snakes and ladders and rolls a dice. What is the probability that she gets a prime number ?

(A) $\frac{5}{6}$ (B) $\frac{3}{6}$

(C) $\frac{1}{2}$ (D) $\frac{1}{6}$

44. Which of the below is correct ?

(A) $31 - 39 = 39 - 31$ (B) $31 \times 39 = 31 + 39$

(C) $31 + 130 = 30 + 31$ (D) $31 + 39 = 7 \times 10$

45. Mrs. G ordered bead boxes in bulk for her craft work shop. The table below shows the total number of boxes and the number of beads it contains

Box#	Number of beads
1	52
2	104
3	156
4	208

Based on the given data in the table, what will be the total number of beads in 10 boxes ?

(A) 520 (B) 1520

(C) 260 (D) 1254

46. Which of the below statements is correct ?

(A) 13 × 19 = 16 × 12

(B) 12 × 17 = 17 + 15

(C) 13 × 19 = 31 + 7

(D) 13 × 19 = 19 × 13

47. Line segment PQ is marked using the rule "add 2" to mark points in a pattern on the line. Based on the rule what will be point A marked as ?

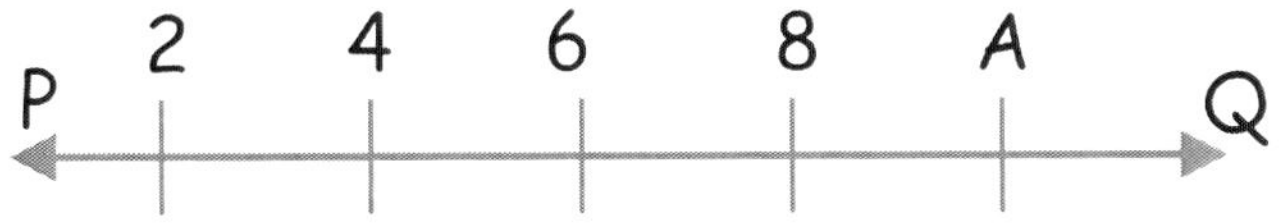

(A) 16

(B) 12

(C) 10

(D) 18

48. Which of the below is correct ?

(A) (22 × 27) - 21 = 225 - (27 × 21)

(B) $\frac{(22 \times 27)}{21} = \frac{22}{(27 \times 21)}$

(C) (22 × 27) + 21 = 22 + (27 × 21)

(D) (22 × 27) × 21 = 22 × (27 × 21)

49. Which of the following is a reflection ?

(A) 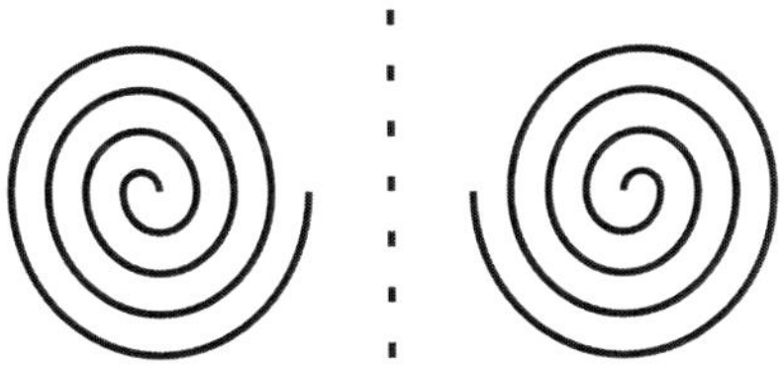

(B) 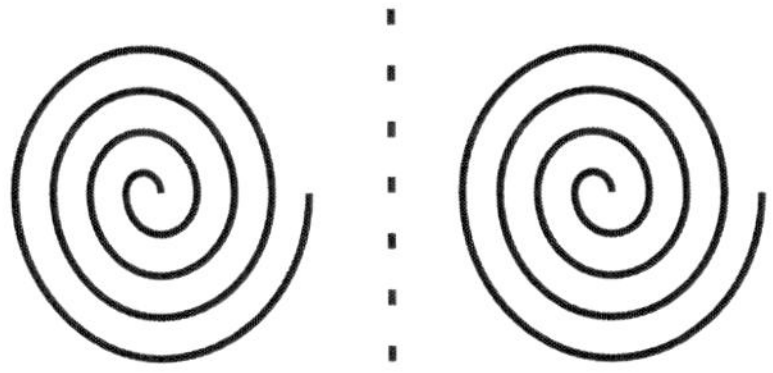

(C) 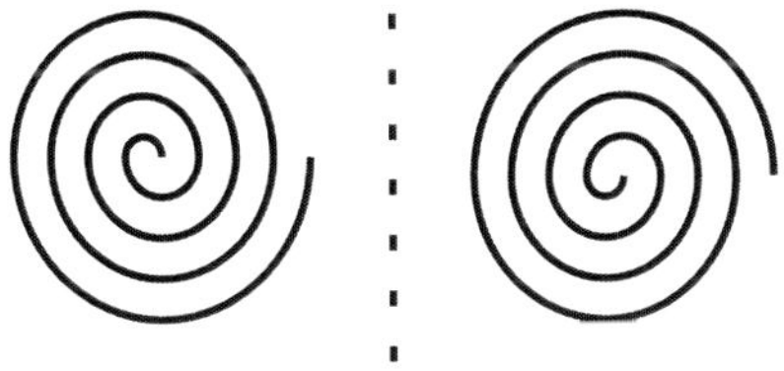

(D) 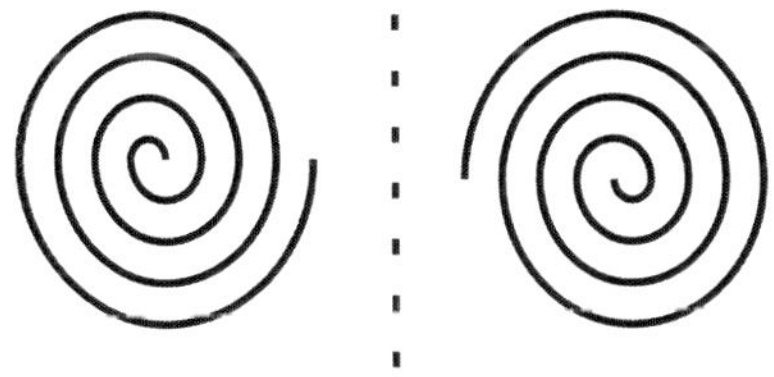

50. The points on this number line represent a pattern.

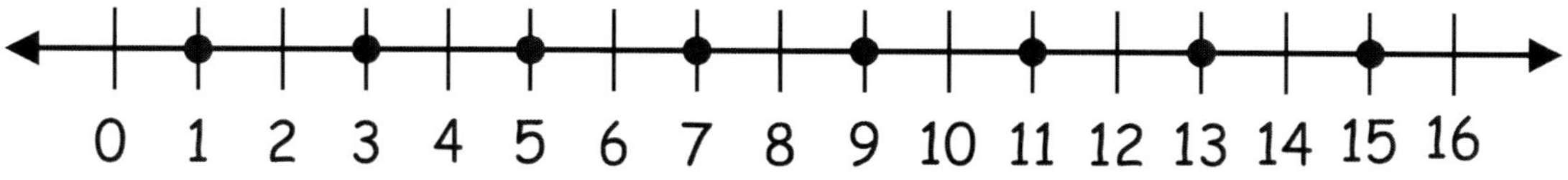

Which of these number lines continues this pattern of points in the same way ?

(A)

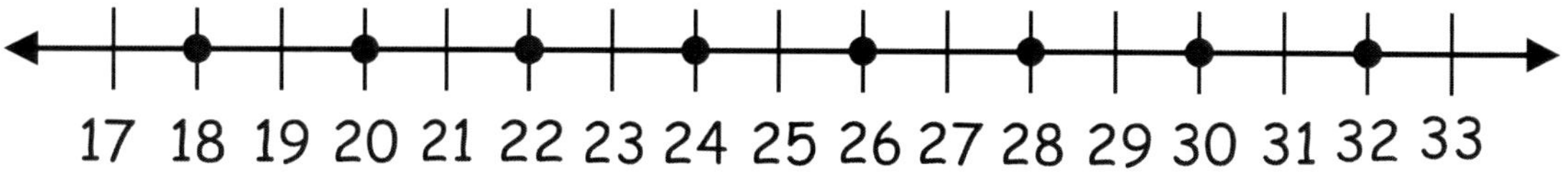

(B)

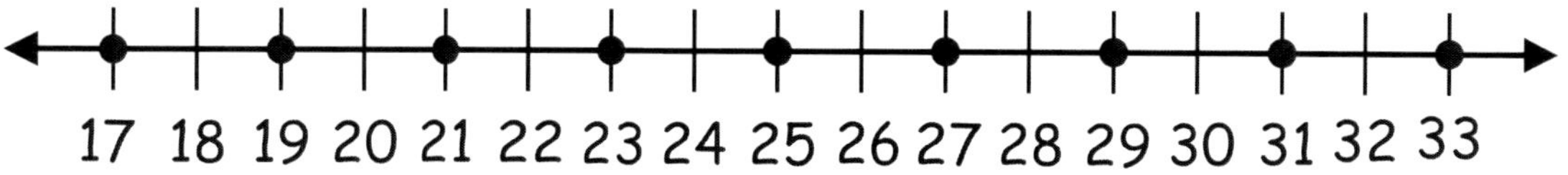

(C)

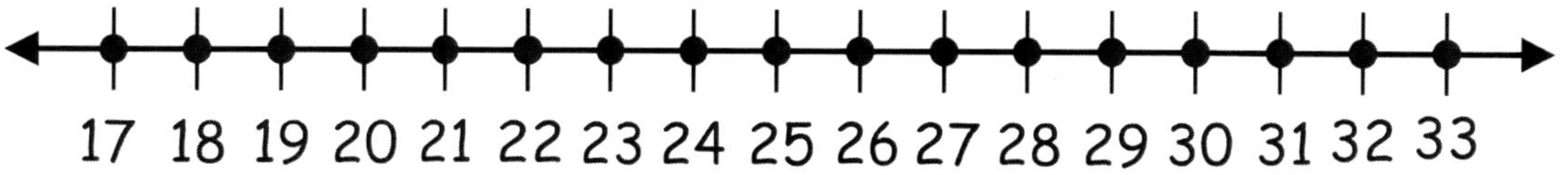

(D)

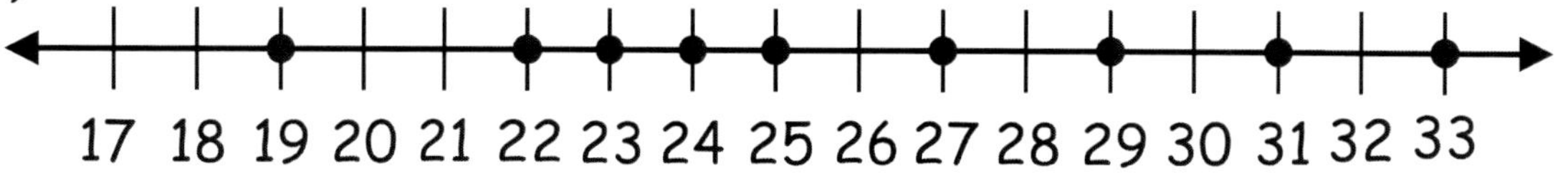

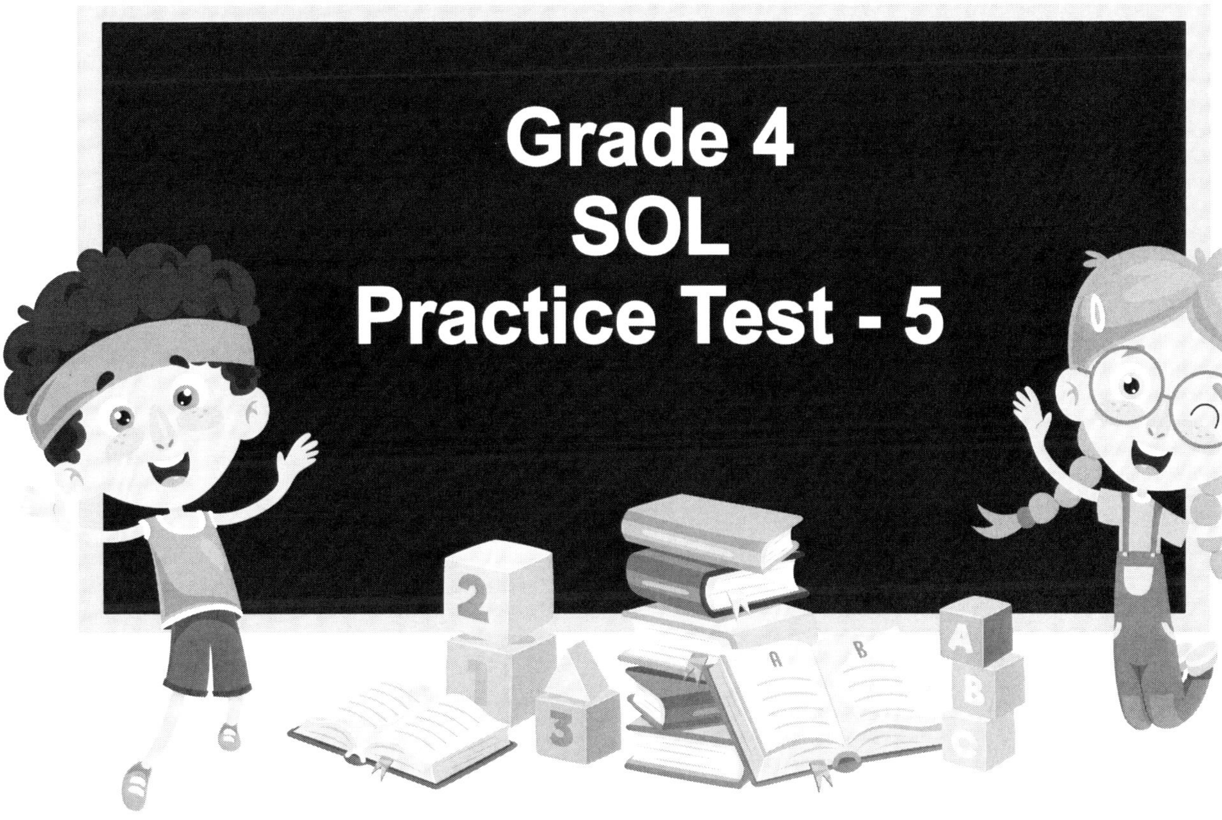
Grade 4
SOL
Practice Test - 5

1. Subtract

$$\begin{array}{r} 789{,}324 \\ -\ 15{,}345 \\ \hline \end{array}$$

(A) 773,969 (B) 777,989

(C) 775,979 (D) 773,979

2. Which of the below product is between 350 and 400 ?

(A) 32 × 12 (B) 35 × 12

(C) 47 × 12 (D) 43 × 12

3. Simplify $\frac{1}{3} + \frac{5}{6} = \boxed{?}$

(A) $\frac{7}{3}$ (B) $\frac{5}{3}$

(C) $\frac{7}{6}$ (D) $\frac{11}{3}$

4. A milk pot contains 4.56 liters of milk. Jose adds 1.25 liters of water to it. How many litres of milk is in the pot now ?

(A) 5.81 (B) 5.71

(C) 6.81 (D) 5.82

5. Below model is equal to 1.

Each of the below model is shaded to represent a decimal number

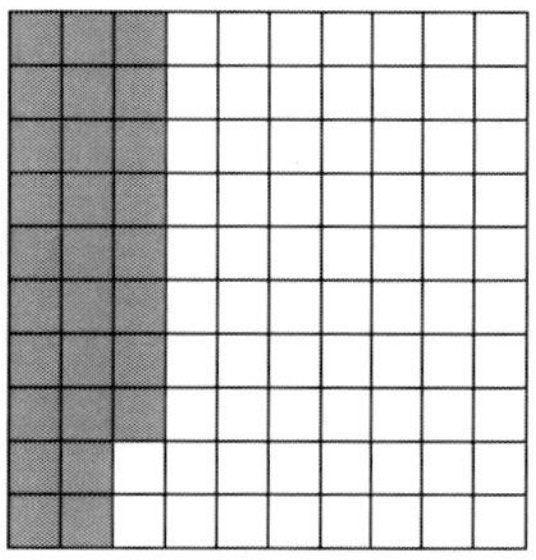

(A) 0.27

(B) 0.28

(C) 0.29

(D) 0.23

6. Which of the below statements is correct ?

(A) $\frac{88}{8} = 6 - 5$

(B) $\frac{88}{8} = 6 + 5$

(C) $\frac{88}{11} = \frac{6}{2}$

(D) $\frac{88}{11} = 88 + 11$

7. Evaluate 45.23 - 0.95.

(A) 44.25

(B) 44.27

(C) 44.25

(D) 44.28

8. Simplify $\frac{643}{6} = \boxed{?}$

(A) 106 R 1

(B) 107 R 1

(C) 107

(D) 107 R 2

9. Which best describes the difference 5,621 - 1111 ?

(A) Closer to 4500 than 5000

(B) Closer to 4000 than 5000

(C) Closer to 3500 than 4000

(D) Closer to 3500 than 4500

10. Simplify $\frac{770}{11} = \boxed{?}$

(A) 70

(B) 68

(C) 71

(D) 77

11. Find the product of 45 and 32

(A) 144

(B) 146

(C) 1440

(D) 1450

12. Simplify $\frac{127}{3} = \boxed{?}$

(A) 41 R 0

(B) 41 R 1

(C) 42 R 1

(D) 42 R 2

13. Which two factors have a product of 675 ?

(A) 25 × 26

(B) 25 × 27

(C) 26 × 23

(D) 26 × 27

14. Round 34,932 to the nearest thousand.

(A) 34,000

(B) 34,900

(C) 35,000

(D) 34,930

15. Write "Eight and six hundred thirty-eight thousandths" in standard form.

(A) 8.0638

(B) 8.6038

(C) 86.038

(D) 8.638

16. Which of the following shows a fraction model and a decimal model that both represent the same value ?

(A) 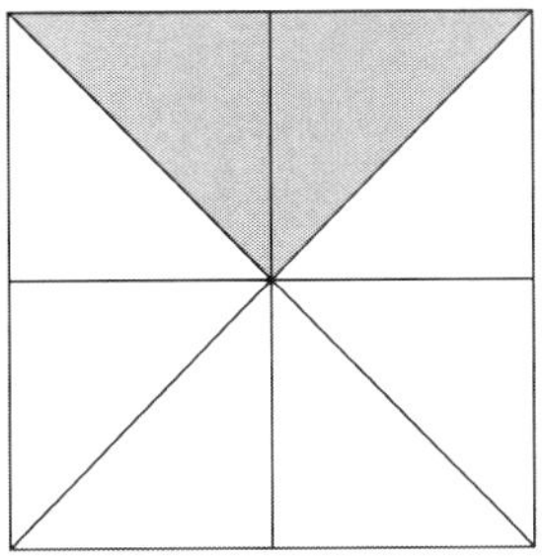 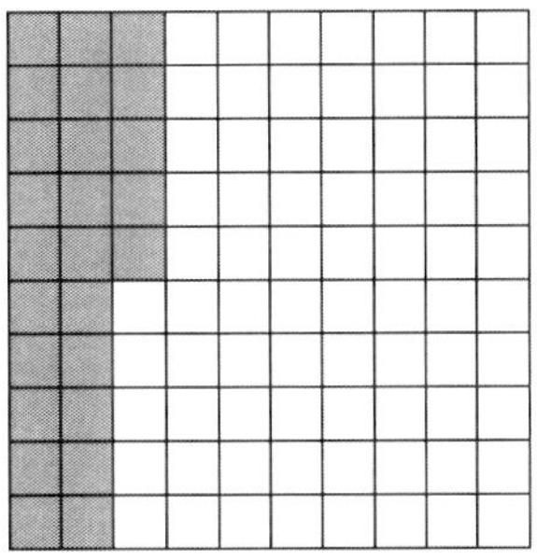

(B) 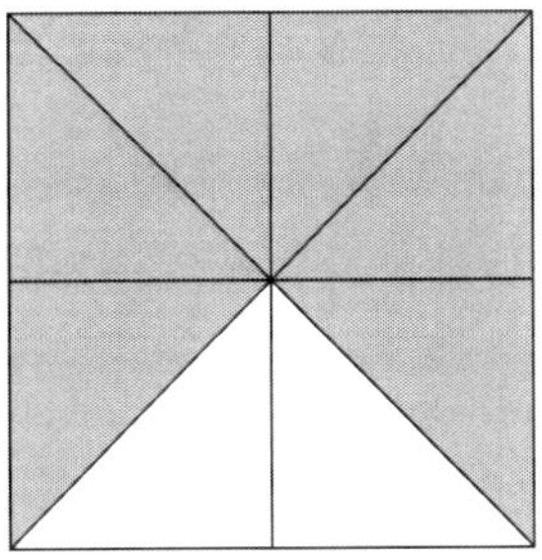 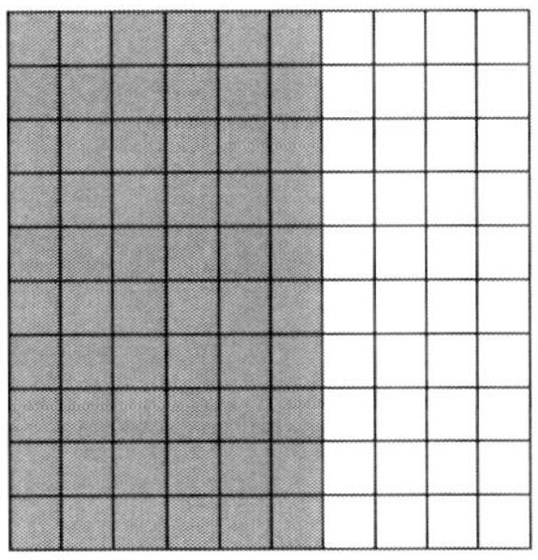

(C) 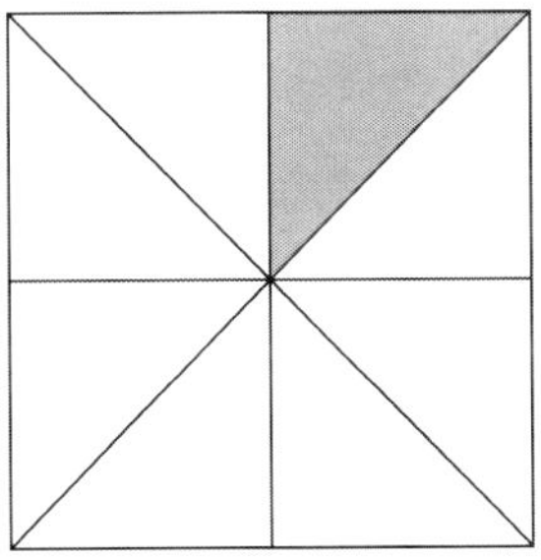 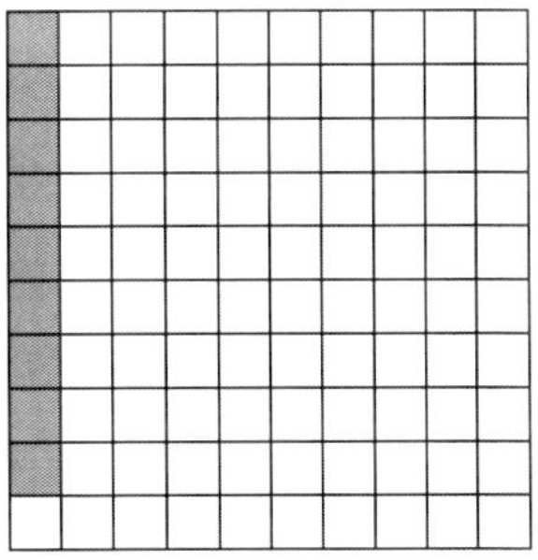

(D) 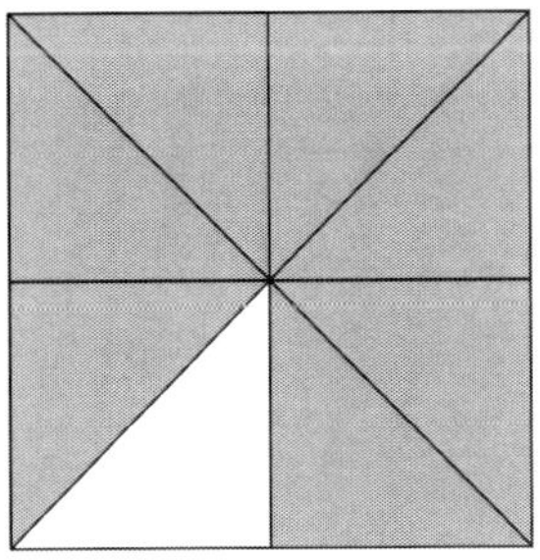 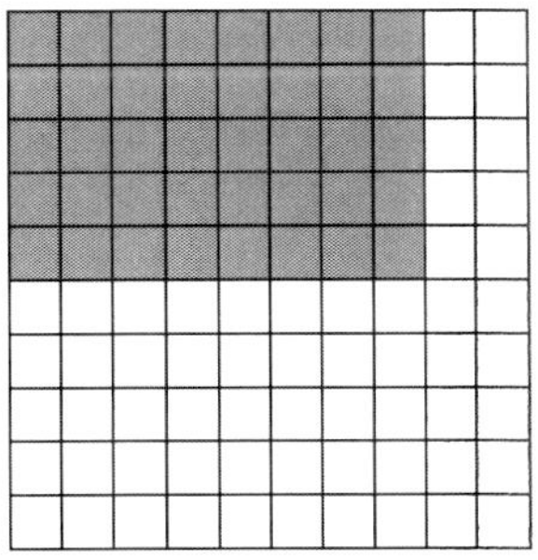

17. Which of the below model is greater than $\frac{3}{5}$

(A) 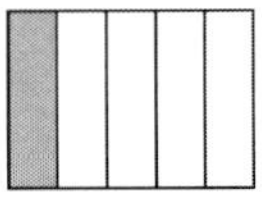

(B) 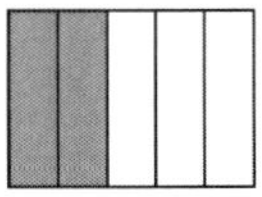

(C) 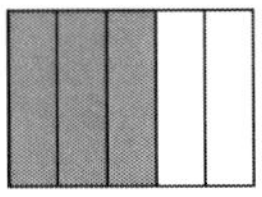

(D) 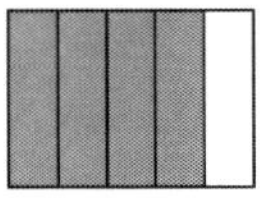

18. Which of the below is a true statement ?

(A) 334323 > 334321

(B) 324323 < 324313

(C) 323323 = 323333

(D) 324323 > 324531

19. Point X is located at $\frac{2}{3}$ on the number line

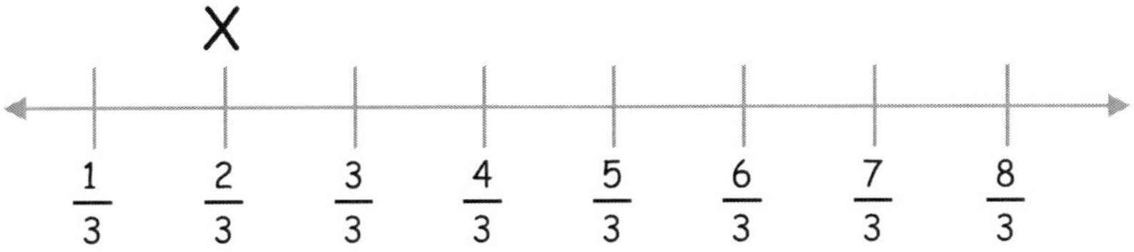

Which of the following is another name for location of X ?

(A) $\frac{4}{5}$

(B) $\frac{3}{7}$

(C) $\frac{8}{12}$

(D) $\frac{5}{9}$

20. Which of the below statements is true ?

(A) 0.704 > 0.733

(B) 0.704 < 0.703

(C) 0.704 > 0.703

(D) 0.704 = 0.740

21. The shaded sections of these models represent two decimal numbers.

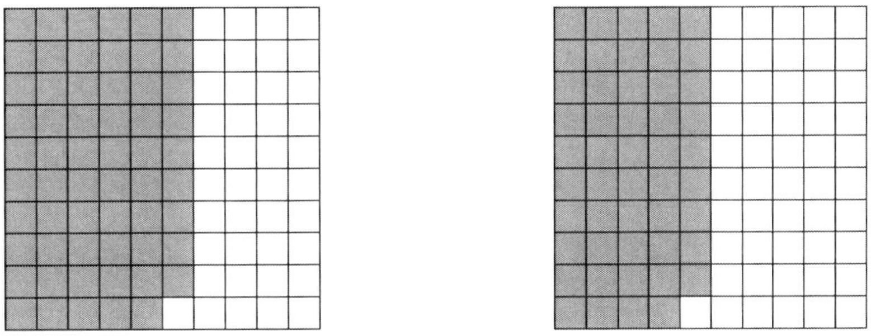

Based on the above which of the below is true ?

(A) 0.59 > 0.49

(B) 0.49 = 0.59

(C) 0.59 < 0.49

(D) 0.59 < 0.44

22. Which if the below shows a semicircle ?

(A) 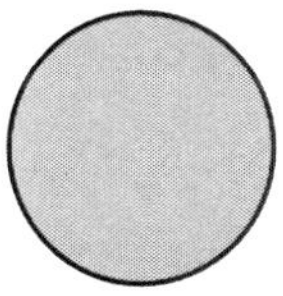

(B) 

(C) 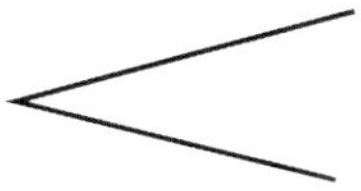

(D) 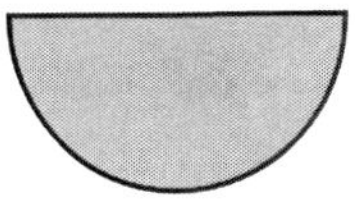

23. Which of the following is closest to the height of the flag post ?

(A) 10 units

(B) 8 units

(C) 4 units

(D) 5 units

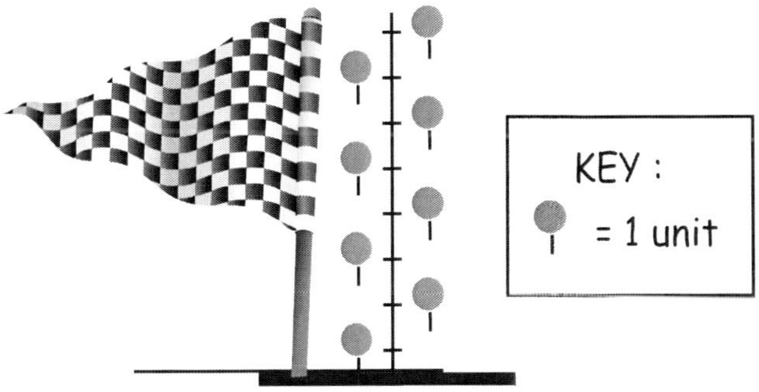

24. Kate is making dessert that requires 16 ounces of milk. How many pounds of milk does she need ?

(A) 1.5 pounds
(B) 3 pounds
(C) 1 pound
(D) 2 pounds

25. Patric high school running track is 350 meters long. How long is the running track in centimeters ?

(A) 17500
(B) 35000
(C) 3500
(D) 35

26. Which of the below is a true statement for line segment ?

(A) It has no end points
(B) It extends only to right side
(C) It extends only to left side
(D) It has two end points

27. Which of the below is an example of right angle ?

(A)
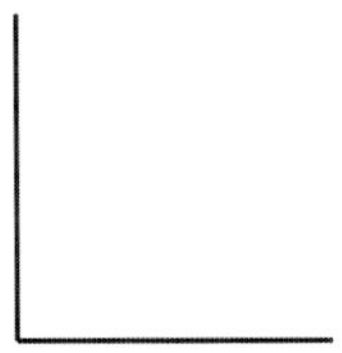

(B)
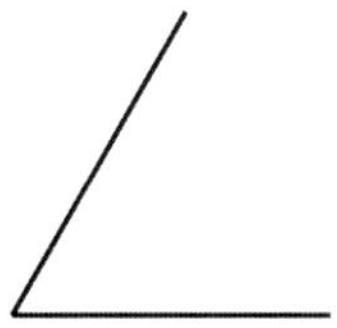

(C)
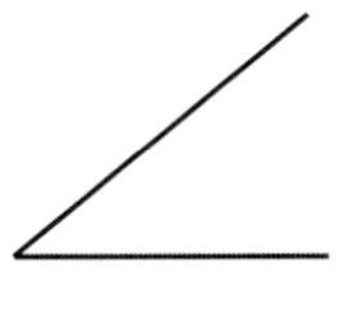

(D)
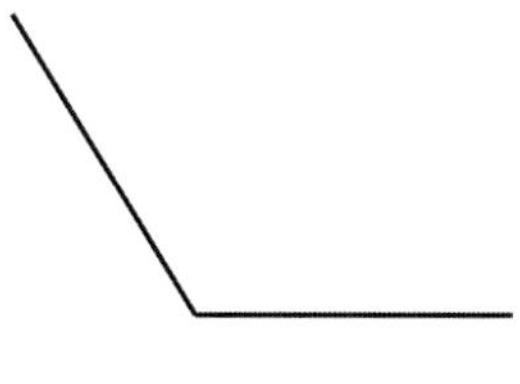

28. Which of the below are congruent.

(A) 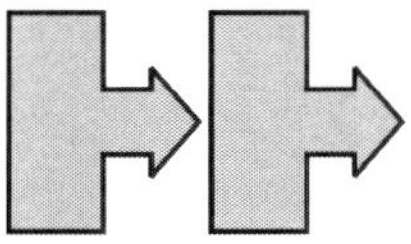

(B) 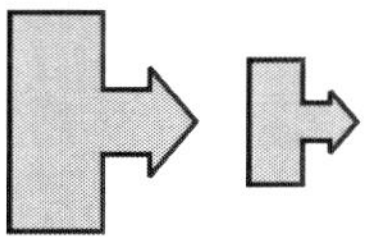

(C) 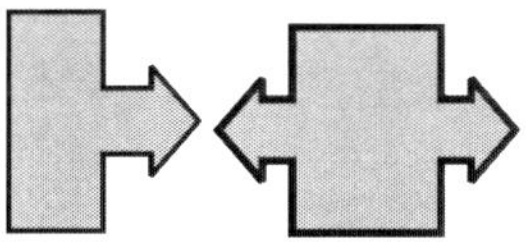

(D) 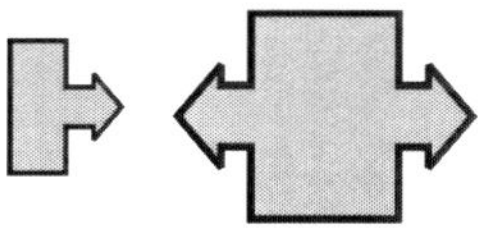

29. Zara is making curry for lunch which requires 1 cup of oil. How much oil does she need in pints ?

(A) 1 pint

(B) 2 pints

(C) Half pint

(D) 1 and half pint

30. Leo purchased a wireless router which supports a range of up to 4,572 cm distance indoors. Calculate that length in meters.

(A) 45720 meters

(B) 45.72 meters

(C) 457200 meters

(D) 0.4572 meters

31. Ruby has 1200 nickel's. Which of the below is equivalent to 1200 nickel's ?

(A) $12.00

(B) $120

(C) $60

(D) $240

32. Which order pair represents the missing coordinate of the below 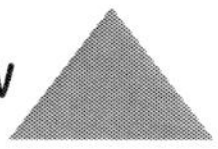 ABC

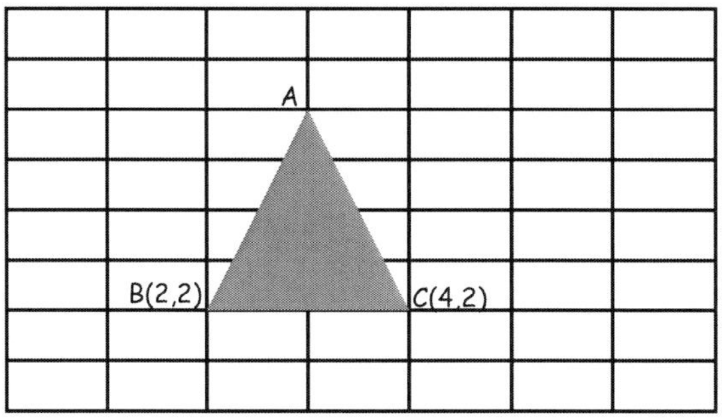

(A) (3,6) (B) (6,6)

(C) (3,3) (D) (6,3)

33. Look at the below graph and find the transformation.

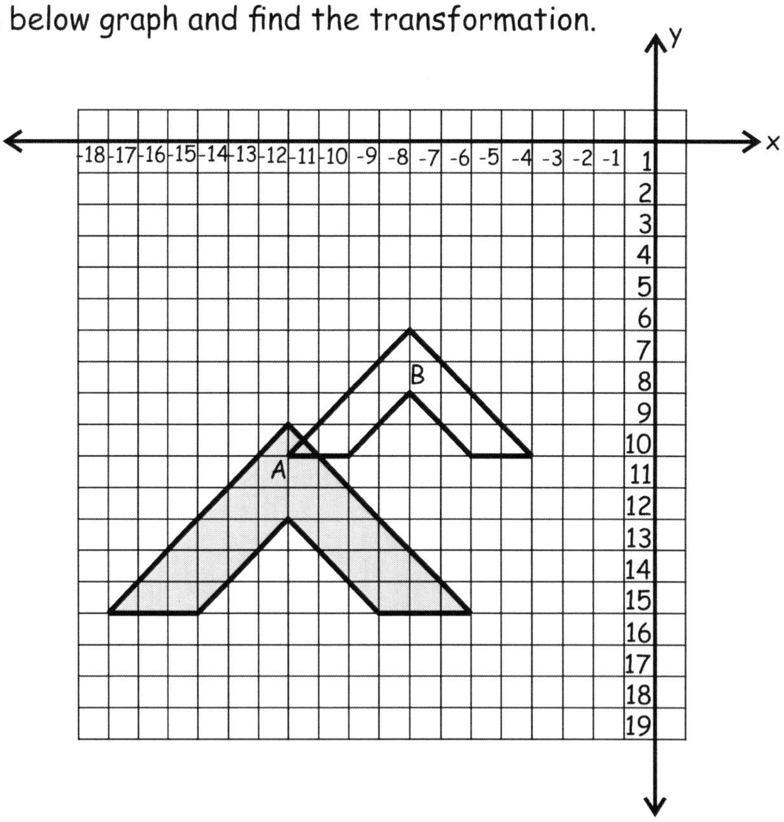

(A) Rotation (B) Symmetry

(C) Reflection (D) Dilation

Two groups of kids were asked to vote for their favorite sport. The graph below shows the results.

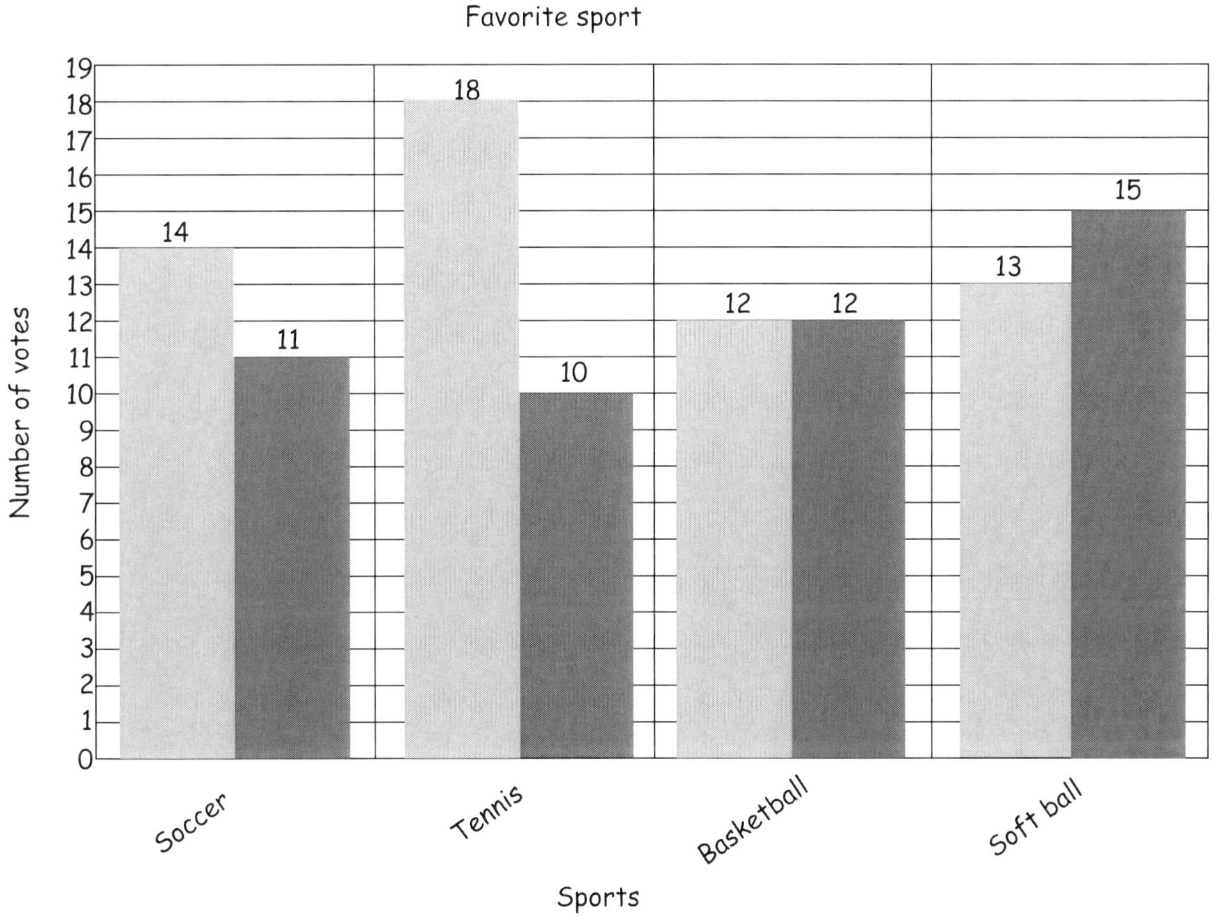

Based on the data in the graph, answer the following questions

34. Which of the following is closest to the total number of kids who liked Soccer as their favorite sport ?

(A) 15
(B) 25
(C) 10
(D) 35

35. How many students voted for tennis ?

(A) 18
(B) 15
(C) 13
(D) 28

36. Magic Machine throws out the below numbers

12 , 11 , 4 , 15 , 17 , 19 , 14 , 16 , 22 , 26 , 32.

George picks a number. What is the probability that the number is an even number ?

(A) $\frac{4}{11}$

(B) $\frac{7}{11}$

(C) $\frac{3}{11}$

(D) $\frac{1}{11}$

37. The table below shows various Christmas gifts that are distributed to children.

Teddy Bear	6
Mickey Mouse	11
Rubik's Cube	2
Connect 4	29

Sam receives one of the above. What is the chance of him getting a Rubik's cube ?

(A) Impossible

(B) Certain

(C) Likely, but not certain

(D) Unlikely, but not impossible

38. Which of the below is the example of reflection ?

(A) 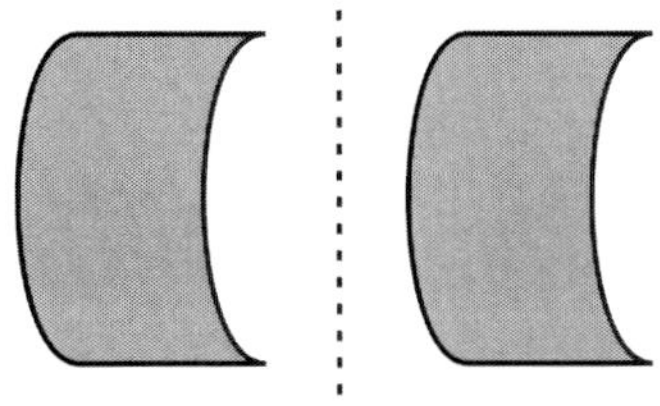

(B) 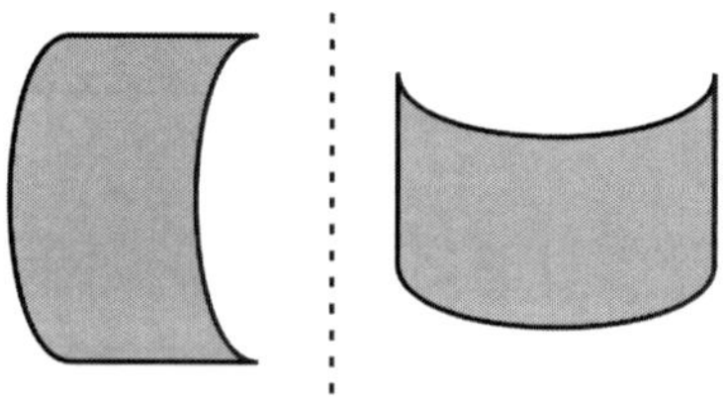

(C) 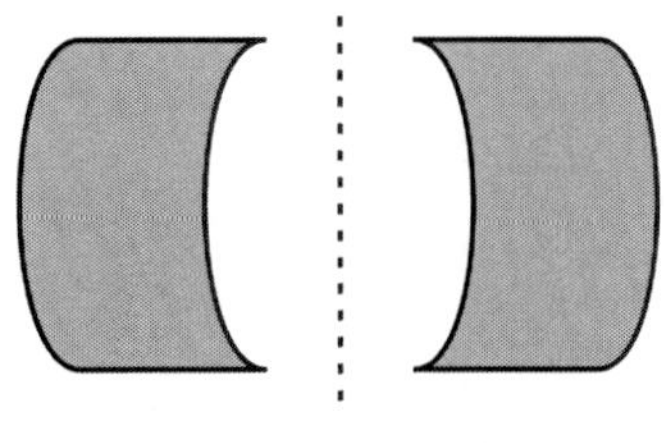

(D) 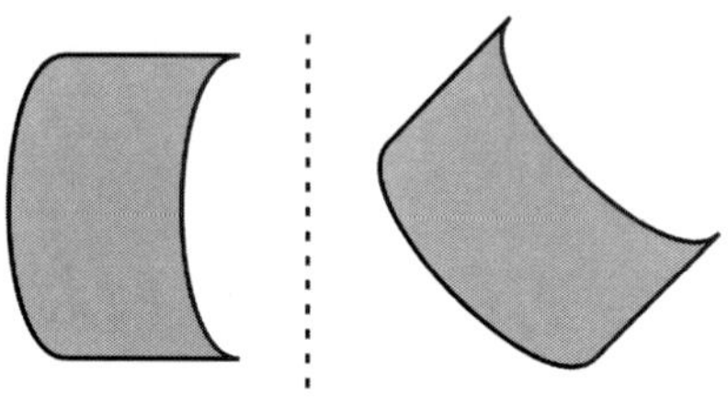

39. Iris will pick one tile from a bag without looking. From which of the following bags is she likely but not certain to pick a tile with a "35" on it ?

(A)

4	3	11	4
9	11	35	7
10	35	8	9
11	4	1	3

(B)

35	35	35	35
35	35	35	35
10	7	8	9
11	4	10	3

(C)

11	3	11	4
8	11	1	11
11	7	11	9
11	4	11	11

(D)

4	3	11	35
8	35	1	4
10	7	8	11
11	35	10	3

40. Grace is playing monopoly and rolls a dice. What is the probability that she gets an even number ?

(A) $\frac{1}{2}$

(B) $\frac{1}{6}$

(C) $\frac{1}{3}$

(D) $\frac{5}{6}$

41. Which of the below is correct ?

(A) 11 - 10 = 10 - 11

(B) $\frac{11}{10} = 11 \times 10$

(C) 10 + 10 = 10 × 10

(D) 11 + 10 = 10 + 11

42. Mr. K ordered Science kits that contain circuits for the Science experiments. The table below shows the total number of kits and the circuits it contains.

Box#	Number of Circuits
1	6
2	12
3	18
4	24

Based on the given data in the table, what will be the total number of circuits in 11 kits ?

(A) 66

(B) 62

(C) 68

(D) 60

43. Which of the below statements is correct ?

(A) $3 \times 7 = 5 \times 11$

(B) $3 \times 7 = 7 + 3$

(C) $3 \times 7 = 20 + 6$

(D) $3 \times 6 = 9 \times 2$

44. Line segment PQ is marked using the rule "add 7" to mark points in a pattern on the line. Based on the rule what will be point A marked as ?

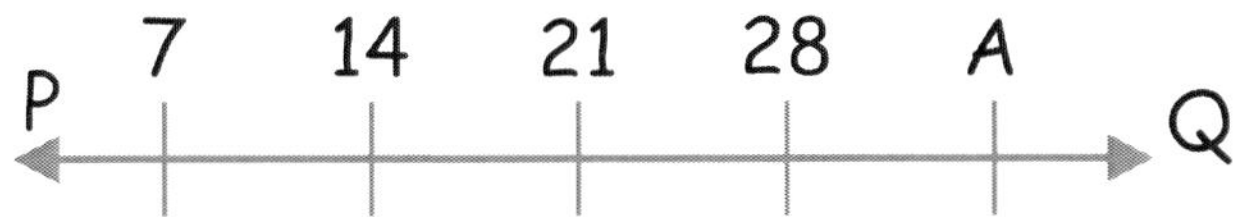

(A) 25

(B) 30

(C) 20

(D) 35

45. Which of the below statements is correct ?

(A) $(5 \times 7) \times 11 = 5 \times (7 \times 11)$

(B) $(5 \times 7) + 11 = 5 + (7 \times 11)$

(C) $\frac{(5 \times 7)}{11} = \frac{5}{(7 \times 11)}$

(D) $(5 \times 7) - 11 = 5 - (7 \times 11)$

46. A Gui-Gui magic bag has 10 red beads, 16 yellow beads, 5 white beads, 1 orange beads, 4 black bead. Aria picks one bead from the bag without looking in. Which color bead is least likely to be picked up ?

(A) Red

(B) Yellow

(C) Orange

(D) White

47. The points on this number line represent a pattern.

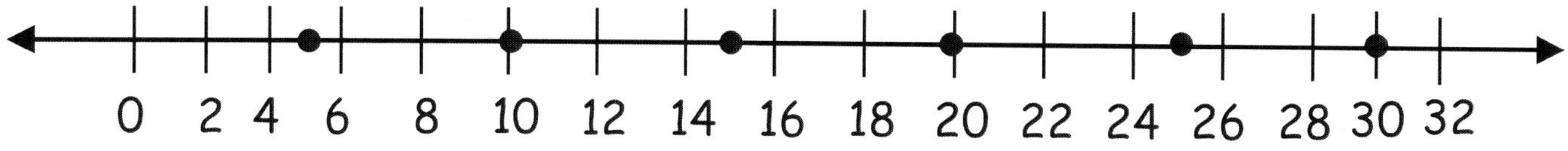

Which of these number lines continues this pattern of points in the same way ?

(A)

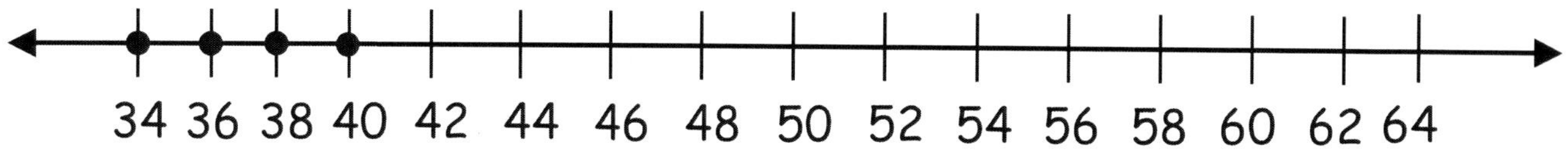

(B)

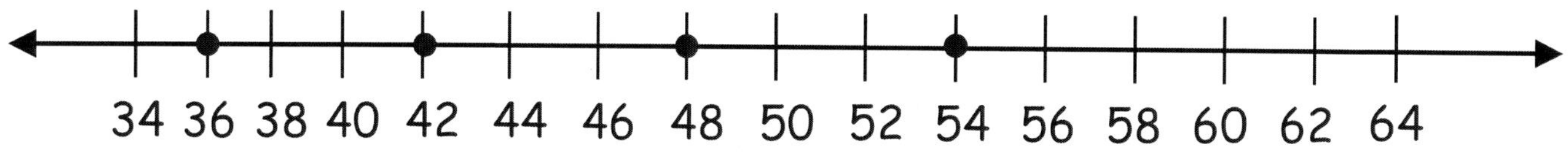

(C)

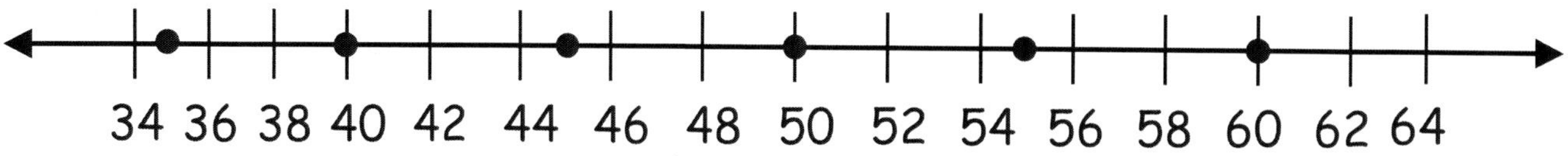

(D)

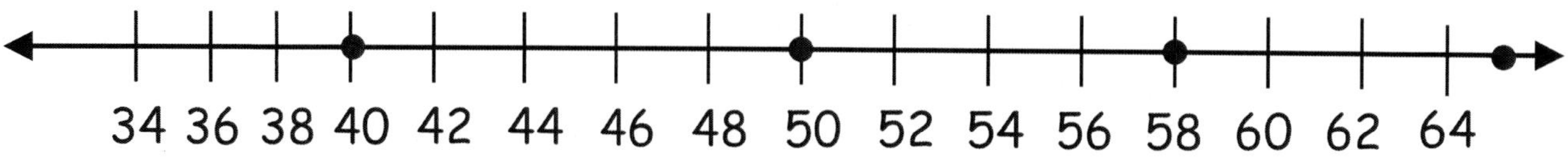

48. Find the next three numbers in the below pattern.

97 , 92 , 87 , 82 , 77 , ...

(A) 72, 67, 62

(B) 72, 62, 57

(C) 72, 67, 57

(D) 67, 62, 57

49. Based on the below table find the rule used on In.

In	Out
5	10
7	14
8	16

(A) Multiply by 2

(B) Add 5

(C) Multiply by 3

(D) Add 7

50. Which of the below statements is correct ?

(A) $14 \times 15 = \frac{190}{2}$

(B) $14 \times 16 = 180 - 44$

(C) $14 \times 16 = 44 + 180$

(D) $14 \times 15 = 14 + 15$

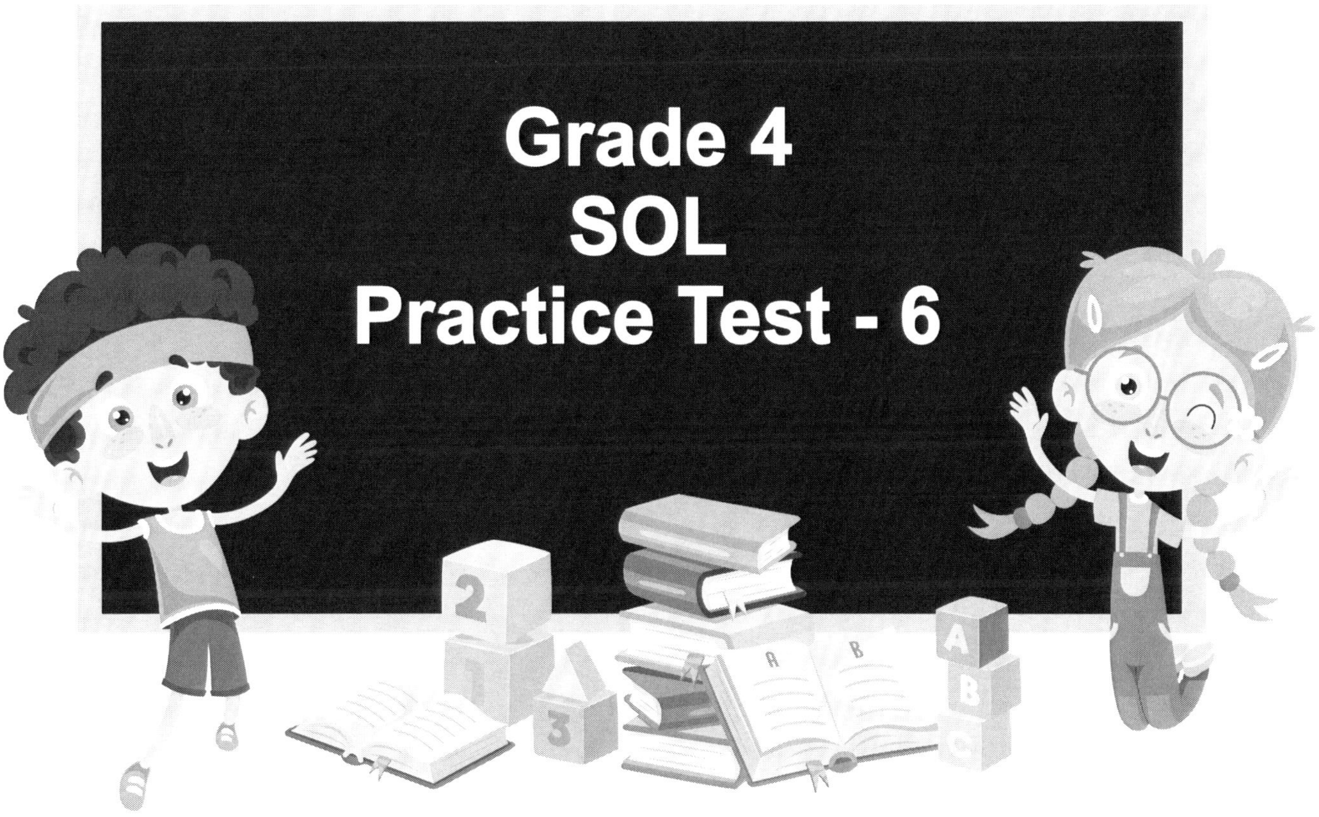
Grade 4
SOL
Practice Test - 6

1. Subtract

$$\begin{array}{r} 93{,}431 \\ -\ 29{,}361 \\ \hline \end{array}$$

(A) 64,070
(B) 64,077
(C) 65,070
(D) 64,707

2. Crazy city has $\frac{3}{8}$ of its residents as males.

How many Crazy city residents are female ?

(A) $\frac{3}{4}$
(B) $\frac{1}{2}$
(C) $\frac{3}{8}$
(D) $\frac{5}{8}$

3. A zig saw puzzle has 33 pieces. Sam has 4 of those puzzles. How many puzzle pieces does he have in all ?

(A) 33 × 4
(B) 33 + 4
(C) 334
(D) $\frac{33}{4}$

4. Simplify $\frac{7}{11} - \frac{3}{11} = \boxed{?}$

(A) $\frac{5}{11}$
(B) $\frac{2}{11}$
(C) $\frac{4}{11}$
(D) $\frac{1}{11}$

5. Estimate $\frac{91}{6} = \boxed{?}$

(A) 15
(B) 14
(C) 16
(D) 18

6. Evaluate 1.94 + 6.87 = ?

(A) 8.81

(B) 8.89

(C) 8.80

(D) 8.99

7. Evaluate $\frac{165}{11}$ = ?

(A) 13

(B) 16

(C) 15

(D) 14

8. Find the difference between 92607 and 66377

(A) 26230

(B) 26235

(C) 26232

(D) 262234

9. Simplify 297 × 18 = ?

(A) 5366

(B) 5340

(C) 5344

(D) 5346

10. Simplify

$\frac{6}{21} + \frac{3}{7}$ = ?

(A) $\frac{15}{21}$

(B) $\frac{16}{21}$

(C) $\frac{10}{7}$

(D) $\frac{14}{21}$

11. Evaluate

$$\begin{array}{r} 77.81 \\ -\ 33.78 \\ \hline \end{array}$$

(A) 44.44 (B) 44.03

(C) 44.01 (D) 43.03

12. The difference between 835 and 402 is little more than

(A) 440 (B) 430

(C) 450 (D) 400

13. What number makes this number sentence true ?

5 + (11 + 8) = (5 + __) + 8

(A) 8 (B) 13

(C) 5 (D) 11

14. Find the product of 27 and 19

(A) 513 (B) 515

(C) 512 (D) 5121

15. The difference between 105,671 and 99,525 is best described as closest to ______.

(A) 61,456 (B) 6,050

(C) 5,146 (D) 6,150

16. Find the missing number in the below pattern ?

5 , 12 , 19 , 26 , ? , 40

(A) 33 (B) 35

(C) 40 (D) 32

17. Choose the correct statement from the below options

(A) 3 × 9 = 3 × 7 (B) 3 × 9 = 4 × 9

(C) 4 × 6 = 2 × 12 (D) 3 × 8 = 4 × 7

18. If the below pattern continues to six figures, how many blocks are used all together ?

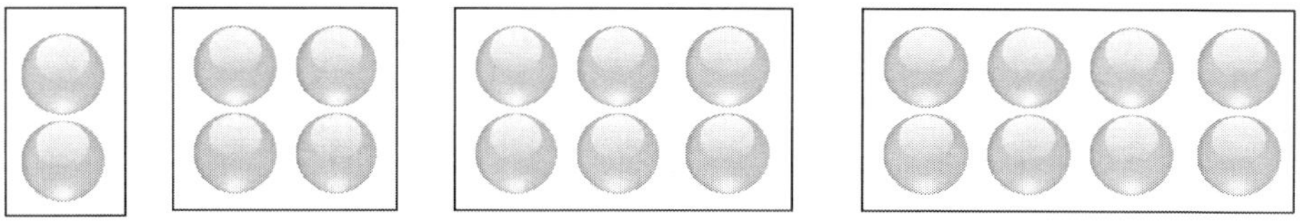

(A) 42 (B) 20

(C) 46 (D) 28

19. Fill the below series by finding the missing number ?

303 , 300 , 297 , 294, ?

(A) 291 (B) 288

(C) 290 (D) 292

20. Which of the below number sentence is not true ?

(A) $2 + (3 + 4) = (2 + 3) + 4$

(B) $\frac{24}{2} = 12 \times 1$

(C) $12 - 2 = 10 - 0$

(D) $(8 \times 3) \times 5 = 8 \times (1 \times 5)$

21. Which number sentence is true ?

(A) $13 + (11 + 6) = 6 + (13 + 11)$

(B) $1 + (2 + 3) = 3 + (2 - 1)$

(C) $11 - (3 + 5) = 5 + (11 + 3)$

(D) $20 + (6 - 3) = 6 + (20 + 3)$

22. Complete the number sentence below

$101 + (102 + 103) = (101 + 102) + ?$

(A) 100

(B) 105

(C) 104

(D) 103

23. Find the pattern that is not correct in the below:

(A) 

(B) 

(C) 

(D) 

24. Below table shows the number of minutes John ran on track in last 4 weeks. If the pattern shown in the table continues in the same way, how many minutes will John practice during the 5th week ?

Running on Track per week	Time in Minutes
1	150
2	175
3	200
4	225

(A) 275 (B) 250

(C) 230 (D) 300

25. A box contains 11 red blocks, 16 blue blocks and 9 silver blocks. Jason picks up a block from the box. Which of the below describes the chances of picking a red block from the box ?

(A) Impossible (B) Unlikely, but not impossible

(C) Likely, but not certain (D) Certain

26. A box contains 11 pencils, 21 erasers, 10 crayons and 6 pens and 1 book marker. Stella picks up a block from the box. What is the probability of picking an eraser ?

(A) $\frac{11}{49}$ (B) $\frac{7}{48}$

(C) $\frac{5}{7}$ (D) $\frac{3}{7}$

27. What is the maximum number of line segments shown in the figure below ?

(A) 16

(B) 11

(C) 18

(D) 21

28. Which of the below is the correct statement for the Square ABCD ?

(A) All sides are not equal

(B) Opposite sides are not parallel

(C) Two obtuse angles

(D) Four right angles

29. Estimate the volume of the juice in the measuring jar.

(A) 8 milliliters

(B) 7 milliliters

(C) 5 milliliters

(D) 10 milliliters

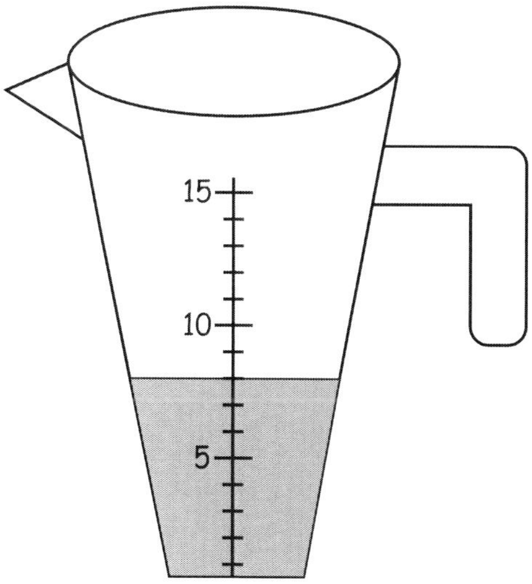

30. A Peacock feather is of the length 3 cm. What is its approximate length in inches ?

(A) 1.2 inches

(B) 1.5 inches

(C) 2 inches

(D) 1.8 inches

31. Which pair of the below lines are parallel ?

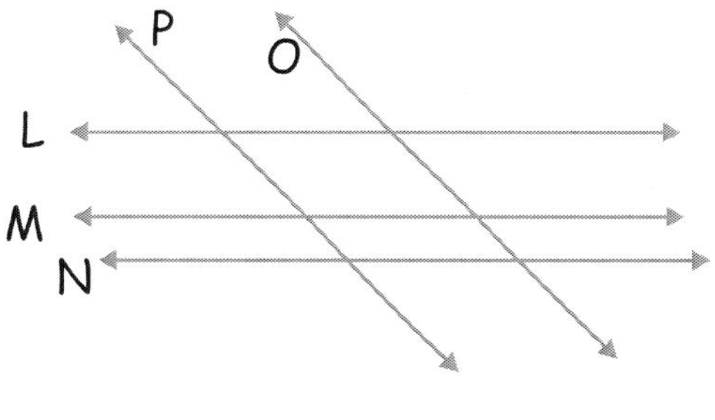

(A) L, M are parallel

(B) M, O are parallel

(C) L, P are parallel

(D) N, O are parallel

32. Newly born kitten weighs 3 pounds. What is the kitten's weight in ounces ?

(A) 32 ounces

(B) 48 ounces

(C) 16 ounces

(D) 80 ounces

33. Which of the below pairs of figures congruent ?

(A) 

(B) 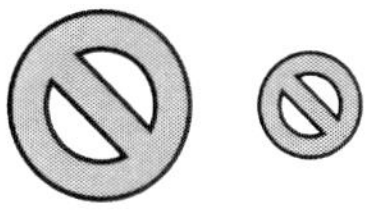

(C) 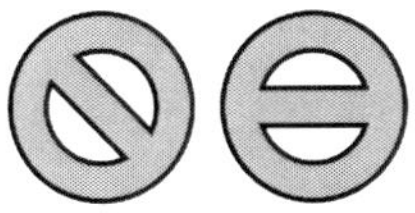

(D) 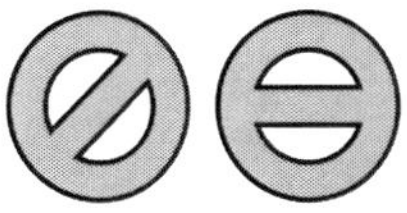

34. Below model represents $\frac{4}{10}$

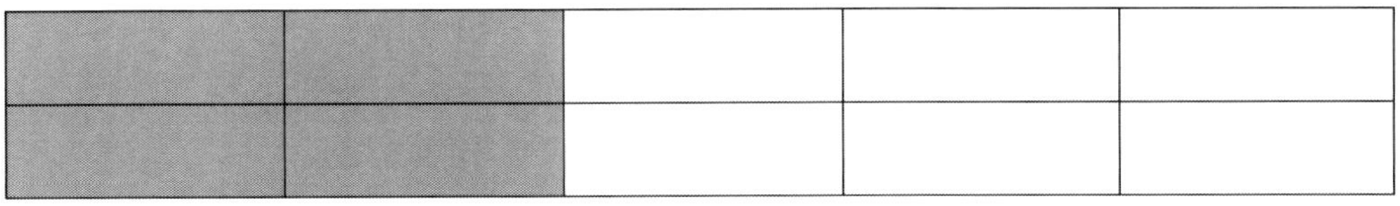

Which of the below is also equal to $\frac{4}{10}$?

(A) 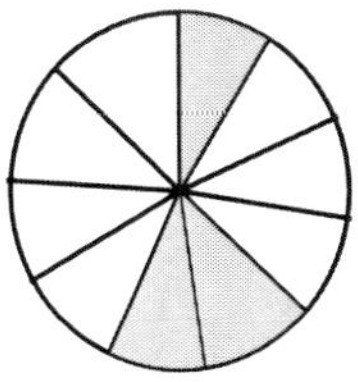

(B) 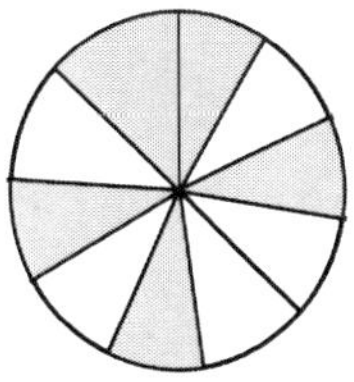

(C) 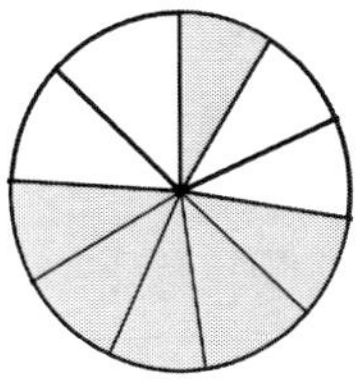

(D) 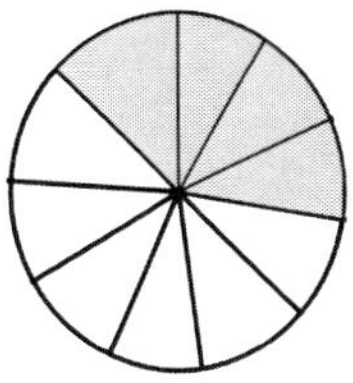

35. If below model equals to 1

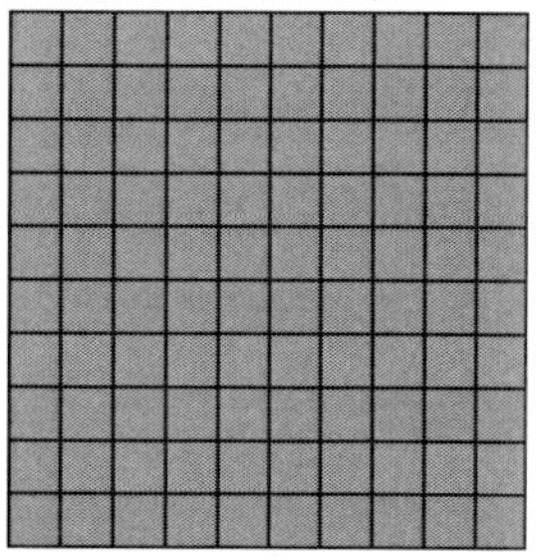

then

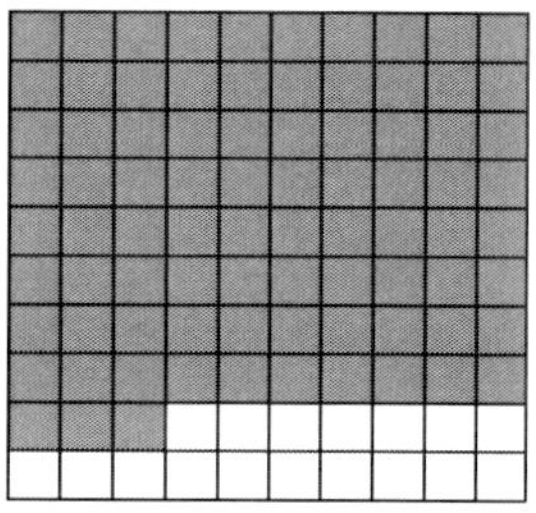

(A) 0.008

(B) 8.03

(C) 0.83

(D) 8.303

36. Which of the below will round to 32 ?

(A) 31.105

(B) 31.99

(C) 31.099

(D) 32.599

37. Use your ruler to answer this question. Estimate the perimeter ?

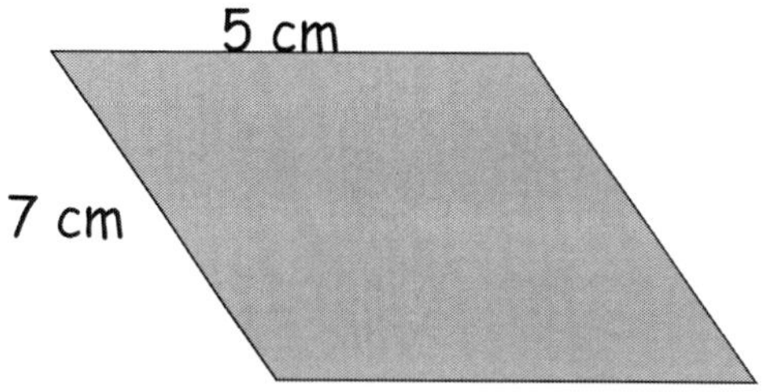

(A) 35 cm^2

(B) 12 cm

(C) 14 cm

(D) 24 cm

38. Which of the below fractions has least value ?

(A) $\frac{1}{5}$ (B) $\frac{1}{2}$

(C) $\frac{1}{6}$ (D) $\frac{1}{3}$

39. Round 357,232 to the nearest ten thousand.

(A) 360,000 (B) 357,000

(C) 400,000 (D) 350,000

40. Round 675,537,232 to the nearest ten million.

(A) 676,000,000 (B) 676,537,232

(C) 675,537,232 (D) 680,000,000

41. Which of the below statement is correct ?

(A) $2003 < 2002$ (B) $9909 < 9988$

(C) $4509 = 5600$ (D) $9909 > 9999$

42. Write 99,999 in words.

(A) Ninety nine thousand, nine hundred and ninety nine.
(B) Ninety thousand nine, nine hundred and nine.
(C) Nine thousand nine, nine hundred and nine.
(D) Nine thousand nine, nine nine and nine

43. Which of the below statements is correct ?

(A) $\frac{3}{9} < \frac{7}{11}$

(B) $\frac{3}{9} > \frac{7}{11}$

(C) $\frac{3}{9} = \frac{7}{11}$

(D) $\frac{5}{9} < \frac{3}{11}$

44. Which ordered pair would have represented the square/parallelogram ?

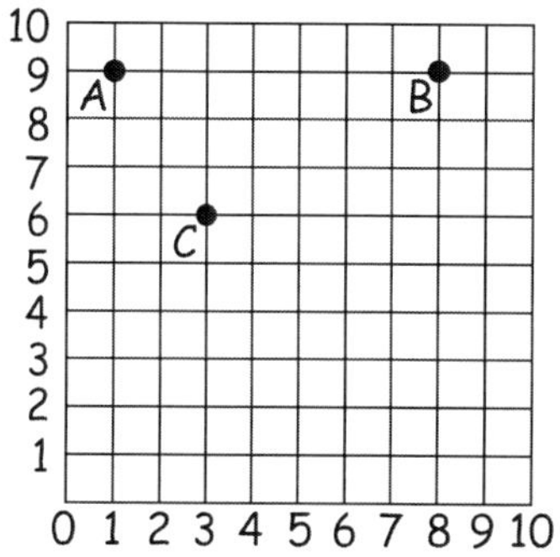

(A) (10,10)

(B) (6,10)

(C) (10,6)

(D) (6,6)

45. Dan measures bananas on the balance scale. Based on the information given below estimate the weight of one banana.

(A) 14

(B) 17

(C) 10

(D) 18

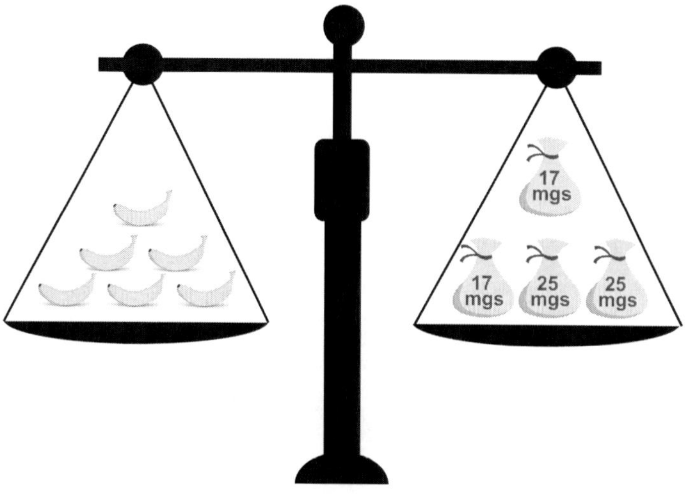

46. Estimate the length of the book shelf ?

(A) 1 unit

(B) 6 units

(C) 8 units

(D) 12 units

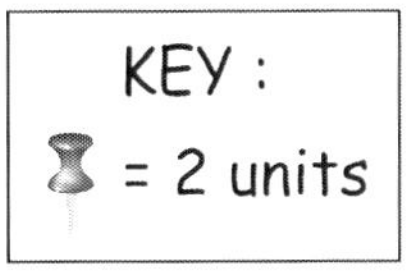

47. Which of the below is an example of reflection ?

(A)
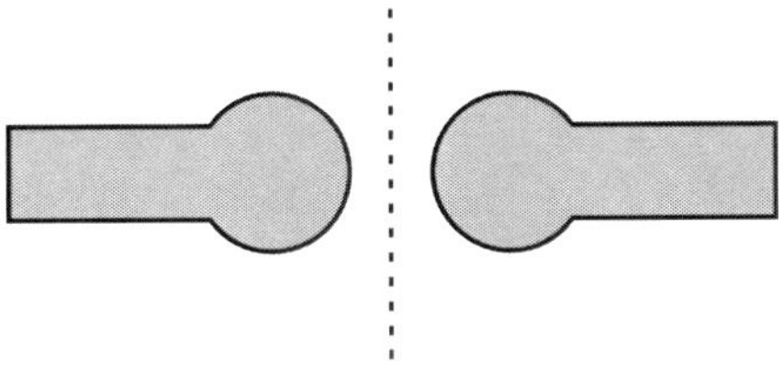

(B)
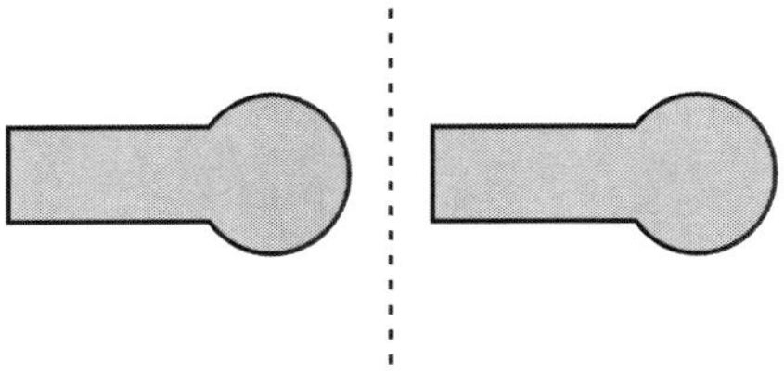

(C)
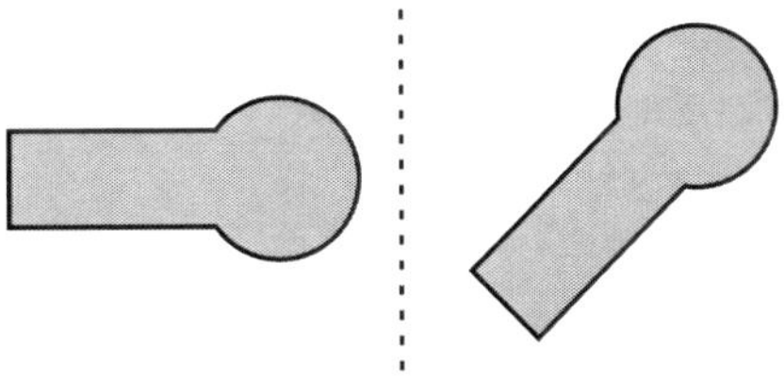

(D)
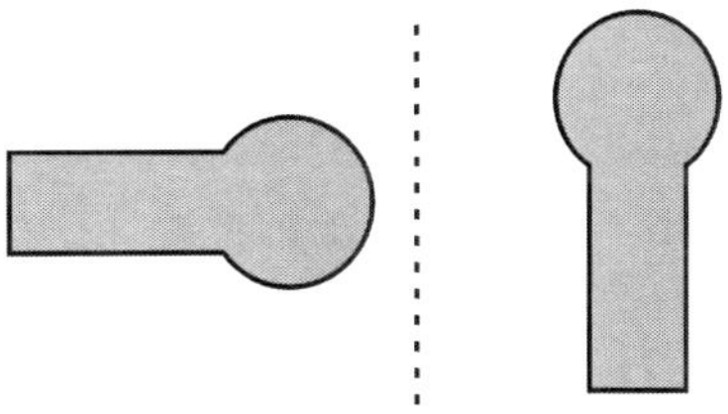

48. Below graph shows the number of puzzles Stella solved over her spring break. Which of the below is the estimated total number of puzzles she solved ?

(A) 40

(B) 45

(C) 50

(D) 55

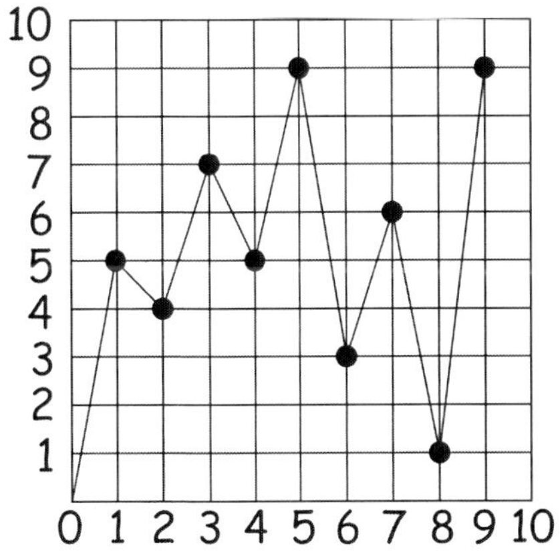

49. Which of the below is an example of acute angle ?

(A) 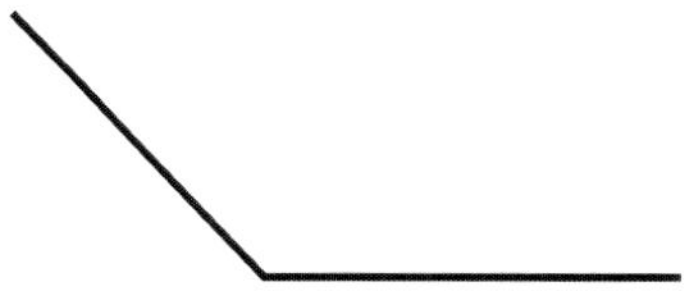

(B) 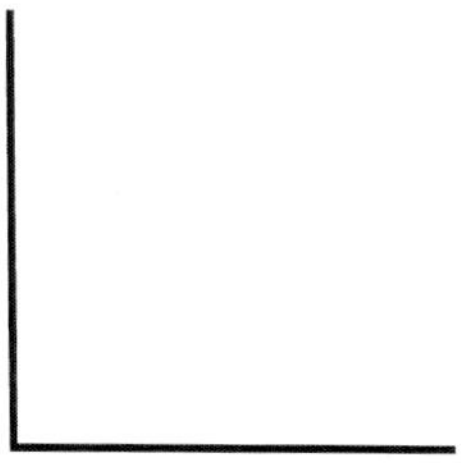

(C)

(D) 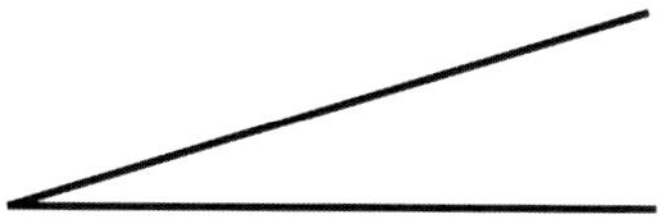

50. Based on the below table find the rule used on In.

In	Out
100	10
120	12
175	17.5

(A) Add 10

(B) Divide by 5

(C) Multiply by 10

(D) Divide by 10

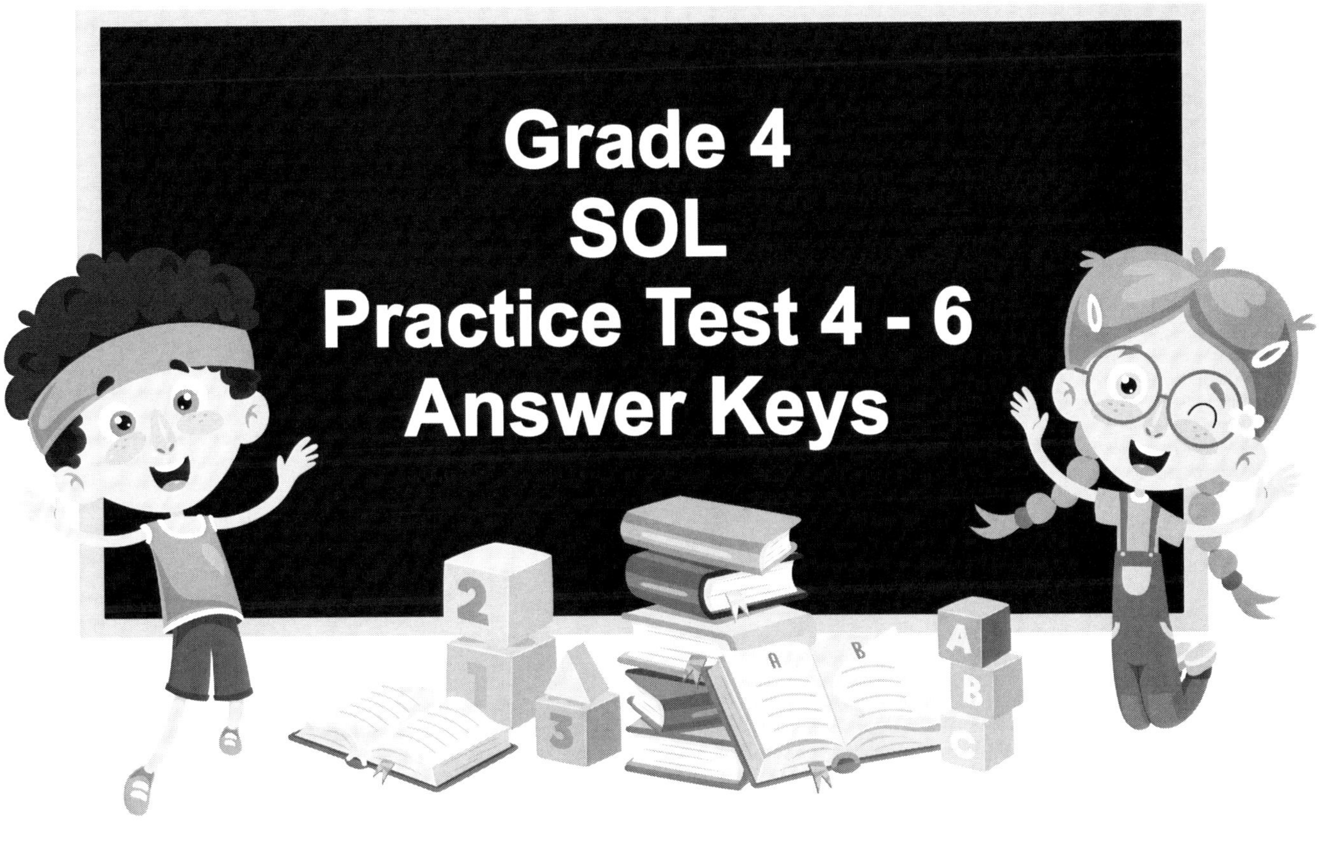
Grade 4
SOL
Practice Test 4 - 6
Answer Keys

Answer Key Test - 4

1. A
2. D
3. A
4. A
5. C
6. A
7. B
8. B
9. B
10. D
11. C
12. A
13. B
14. D
15. B

16. A

17. B

18. D

19. A

20. D

21. C

22. A

23. C

24. A

25. A

26. B

27. A

28. C

29. A

30. C

31. B

32. A

33. A

34. A

35. D

36. A

37. C

38. C

39. A

40. B

41. D

42. A

43. C

44. D

45. A

46. D

47. C

48. D

49. A

50. B

Answer Key Test - 5

1. D
2. A
3. C
4. A
5. B
6. B
7. D
8. B
9. A
10. A
11. C
12. C
13. B
14. C
15. D

16. A

17. D

18. A

19. C

20. C

21. A

22. D

23. B

24. C

25. B

26. D

27. A

28. A

29. C

30. B

31. D

32. A

33. D

34. B

35. D

36. B

37. D

38. C

39. B

40. A

41. D

42. A

43. D

44. D

45. A

46. C

47. C

48. A

49. A

50. C

Answer Key Test - 6

1. A
2. D
3. A
4. C
5. A
6. A
7. C
8. A
9. D
10. A
11. B
12. B
13. D
14. A
15. D

16. A

17. C

18. A

19. A

20. D

21. A

22. D

23. B

24. B

25. C

26. D

27. C

28. D

29. A

30. D

31. A

32. B

33. A

34. D

35. C

36. B

37. D

38. C

39. A

40. D

41. B

42. A

43. A

44. C

45. A

46. D

47. A

48. C

49. D

50. D

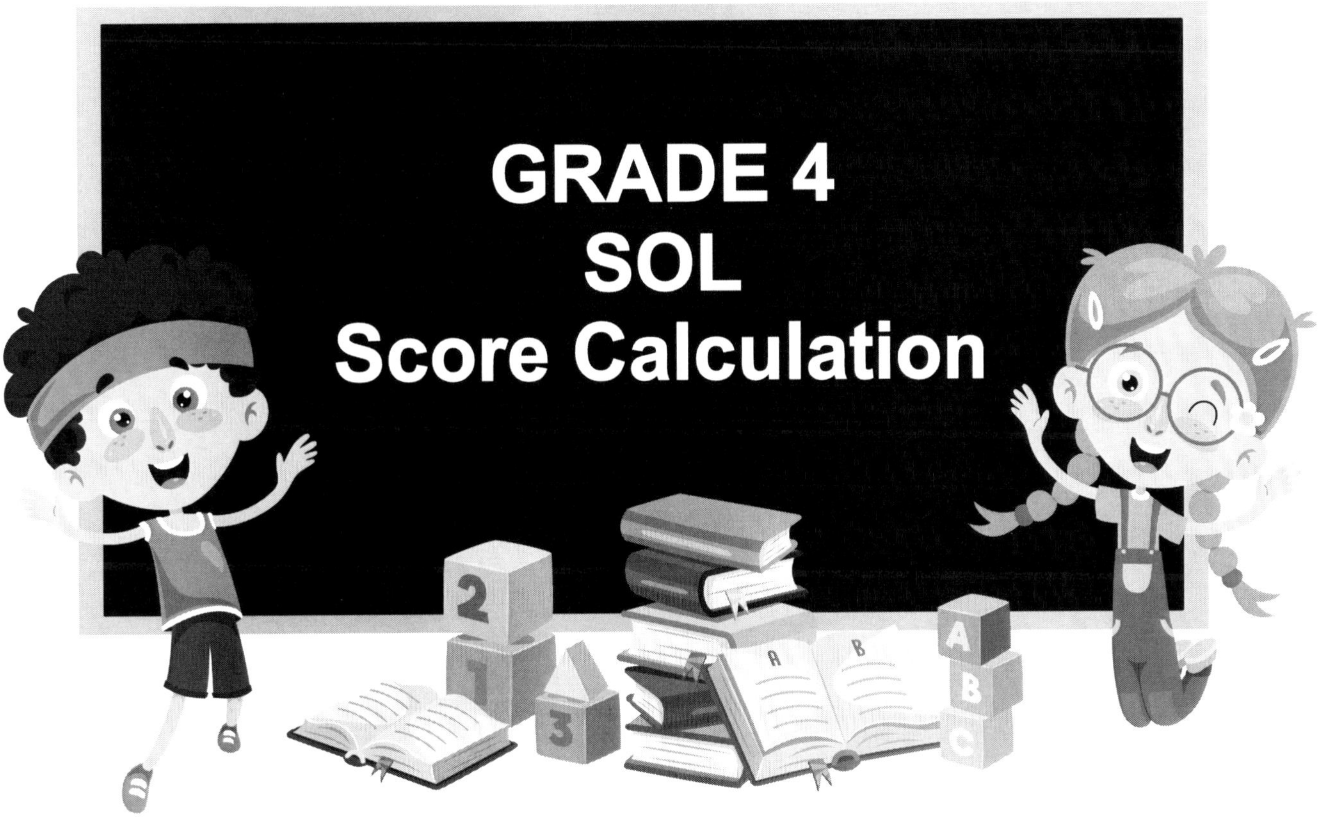
GRADE 4
SOL
Score Calculation

Score calculation

If you get this many times correct :	Then your converted scale scor is :
0	000
1	217
2	249
3	268
4	282
5	293
6	303
7	311
8	319
9	325
10	331
11	337
12	343
13	348
14	353
15	357
16	362
17	366
18	371
19	375

20	379
21	383
22	387
23	391
24	395
25	399
26	403
27	407
28	411
29	415
30	419
31	423
32	427
33	432
34	436
35	441
36	445
37	450
38	456
39	461

SCORE

40	467
41	473
42	480
43	487
44	496
45	505
46	516
47	531
48	550
49	582
50	600

Made in United States
North Haven, CT
05 November 2026

21769181R00050